Technology Transfer for Renewable Energy

The Sustainable Development Programme is the new name (from February 2002) for the Energy and Environment Programme at Chatham House. The Programme works with business, government, and academic and NGO experts to carry out and publish research and stimulate debate across a wide variety of energy, environment and business topics with international implications, particularly those just emerging into the consciousness of policy-makers. Research by the Programme is supported by generous contributions of finance and technical advice from the following organizations:

Amerada Hess Ltd
Anglo American plc
BG Group
BP Amoco plc
British Energy plc
British Nuclear Fuels plc
Department for the Environment, Food and Rural Affairs (UK)
Department of Trade & Industry (UK)
ExxonMobil
Foreign & Commonwealth Office (UK)
Osaka Gas Co. Ltd
Powergen plc
Saudi Petroleum Overseas Ltd
Shell UK
Statoil
Tokyo Electric Power Co. Inc.
TotalFinaElf
TXU Europe Group plc

Technology Transfer for Renewable Energy
Overcoming Barriers in Developing Countries

Gill Wilkins

THE ROYAL INSTITUTE OF INTERNATIONAL AFFAIRS | Sustainable Development Programme

Earthscan Publications Ltd, London

To my mother Freda and to Martin,
for their endless support and encouragement

First published in the UK in 2002 by
The Royal Institute of International Affairs, 10 St James's Square, London SW1Y 4LE
(Charity Registration No. 208223)
and
Earthscan Publications Ltd, 120 Pentonville Road, London N1 9JN

Distributed in North America by
The Brookings Institution, 1775 Massachusetts Avenue NW,
Washington, DC 20036-2188

Copyright © The Royal Institute of International Affairs, 2002

All rights reserved

A catalogue record for this book is available from the British Library.

ISBN 1 85383 753 9 paperback

The Royal Institute of International Affairs is an independent body that promotes the rigorous study of international questions and does not express opinions of its own. The opinions expressed in this publication are the responsibility of the author.

Earthscan Publications Ltd is an editorially independent subsidiary of Kogan Page Ltd and publishes in association with the WWF-UK and the International Institute of Environment and Development.

Typeset by Composition & Design Services, Minsk, Belarus
Printed and bound by Creative Print and Design Wales, Ebbw Vale
Original cover design by Visible Edge
Cover photo by Bill Gillett

Contents

List of figures, tables and boxes

Figures

Tables

Boxes

Abbreviations

A$	Australian dollars
ACE	ASEAN Centre for Energy
ADB	Asian Development Bank
AEAT	AEA Technology
Ah	amp hours
AIJ	activities implemented jointly (pilot JI projects without credits)
AOSIS	Association of Small Island States
AREED	African Rural Energy Enterprise Development
ASEAN	Association of South East Asian Nations
ASTAE	Asia Alternative Energy Programme
AUSAID	Australian overseas aid organization
BANPRES	presidential funds for rural electrification, Indonesia
BOT	build, operate, transfer
BPPT	non-departmental government agency for the assessment and application of technology, Indonesia
BRI	Bank Rakyat Indonesia
Bt	baht (Thai currency unit)
CASE	Centre for Application of Solar Energy, Australia
CB	citizen band (radio)
CDM	Clean Development Mechanism
CEC	Commission of the European Communities
CEE	central and eastern Europe
CERs	certified emissions reductions (for CDM projects)
CHP	combined heat and power
CIDA	Canadian International Development Agency
COP	Conference of the Parties (to the UNFCCC)
CRS	Creditor Reporting System (of the OECD)
CRW	combustible renewables and waste
DAC	Development Assistance Committee (of the OECD)
DANIDA	Danish Agency for International Cooperation
DBP	Development Bank of the Philippines
DC	direct current
DEDP	Department of Energy Development and Promotion, Thailand
DG	Directorate-General (of the European Commission)
DGEED	Directorate General for Electricity and Energy Development (of the Ministry of Mines and Energy), Indonesia

DGIS	Agency for Development and Cooperation in the Netherlands
EC	European Communities
ECA	export credit agency
EDF	European Development Fund
EE	energy efficiency
EFB	empty fruit bunches (oil palm)
EGAT	Electricity Generating Authority of Thailand
EGCO	Electricity Generating Company, Thailand (subsidiary of EGAT)
EIT	economies in transition
ENCON	Energy Conservation Fund, Thailand
ERAP	Energy Resources for the Alleviation of Poverty, Philippines
ERU	emissions reduction unit (for JI projects)
ESCO	energy service company
ESI	electricity supply industry
ESMAP	Energy Sector Management Assistance Programme
EST	environmentally sound technology
EU	European Union
FAO	Food and Agricultural Organization (of the UN)
FDI	foreign direct investment
FFB	fresh fruit bunches (oil palm)
FINESSE	Financing Energy Systems for Small-Scale Energy Users
FPEI	foreign portfolio equity investment
FONDEM	Fondation Energies pour le Monde
FPEI	foreign portfolio equity investment
FSU	former Soviet Union
G7[8]	Group of Seven [Eight] industrialized countries
G77	Group of Seventy-Seven developing countries
GATT	General Agreement on Tariffs and Trade
GDP	gross domestic product
GEF	Global Environment Facility
GHG	greenhouse gas
GNP	gross national product
GOI	Government of Indonesia
GTZ	German Agency for Technology Cooperation
GW	Gigawatt (10^9 watt)
GWh	Gigawatt hour
GWP	global warming potential
IAF	investment advisory facility
IBRD	International Bank for Reconstruction and Development
ICSID	International Centre for Settlement of Investment Disputes (of the World Bank)

IDA	International Development Association (of the World Bank)
IDT	international development target
IEA	International Energy Agency
IEI	International Energy Initiative
IFC	International Finance Corporation
IFI	international finance institution
IOE	Institute of Energy, Vietnam
IP	intellectual property
IPCC	Intergovernmental Panel on Climate Change
IPP	independent power producer
IPR	intellectual property rights
IREDA	Indian Renewable Energy Development Association
JI	joint implementation
JV	joint venture
KEB	Karnataka Electricity Board, India
KUD	village cooperative, Indonesia
kt	kilotonne (10^3 tonnes)
kW	kilowatt (10^3 watt)
kWh_e	kilowatt hour (electricity)
kWp	kilowatt peak
LPG	liquefied petroleum gas
MDG	millennium development goal
MEA	Metropolitan Electricity Authority, Thailand
MIGA	Multilateral Investment Guarantee Agency
MNES	Ministry of Non-conventional Energy Sources, India
MOC	Ministry of Cooperatives and Small Enterprises Development, Indonesia
MOSTE	Ministry of Science, Technology and Environment, Thailand
MoU	memorandum of understanding
Mt	million tonnes (10^6 tonnes)
Mtoe	million tonnes of oil equivalent
MW	megawatt (10^6 watt)
MWp	megawatt (peak)
NEA	National Electrification Administration, Philippines
NEDO	New Energy and Industrial Technology Development Organization of Japan
NEPC	National Energy Policy Committee, Thailand
NEPO	National Energy Policy Office, Thailand
NFFO	Non Fossil Fuel Obligation, United Kingdom
NGO	non-governmental organization
NRE	new and renewable energy

NRSE	new and renewable sources of energy
NSSD	national strategy for sustainable development
O&M	operation and maintenance
OA	official aid (from OECD to EITs)
ODA	official development assistance (from OECD to developing countries)
OECD	Organization for Economic Coordination and Development
PEA	Provincial Electricity Authority, Thailand
PLN	national electricity power generating corporation, Indonesia
PNOC	Philippine National Oil Company
PPA	power purchase agreement
PREP	Pacific Regional Energy Programme
PRESSEA	Promotion of Renewable Energy Sources in Southeast Asia
PRSP	poverty reduction strategy paper
PSKSK	small power purchase agreement, Indonesia
PV	photovoltaic
PVMTI	PV Market Transformation Initiative
QELRC	quantified emissions limitation and reduction commitment
QELRO	quantified emissions limitation and reduction obligation
R&D	research and development
RD&D	research, development and demonstration
RDD&D	research, development, demonstration and deployment
RE	renewable energy
REEF	Global Renewable Energy and Energy Efficiency Fund for Emerging Markets
RES	renewable energy system
RESCO	rural energy service company
RESP	Renewable Energy Small Power project, Indonesia
RET	renewable energy technology
RIIA	Royal Institute of International Affairs
Rp	rupiah (Indonesian currency unit)
Rs	rupee (Indian currency unit)
SADC	Southern African Development Community
SDC	Solar Development Corporation
SDF	Solar Development Foundation
SDG	Solar Development Group
SEA	Southeast Asia
SEB	state electricity board, India
SEC	Solar Energy Company, Kiribati
SEI	Stockholm Environment Institute
SELCO	Solar Electric Light Company
SELF	Solar Electric Light Fund

SEP	Special Energy Programme, Philippines
SHS	solar home systems (household PV application)
SIDA	Swedish International Development Cooperation Agency
SIDS	small island developing states
SME	small and medium-sized enterprise
SNN	sub-sector network
SPIRE	South Pacific Institute for Renewable Energy
SPP	small power producer
TCAPP	Technology Cooperation Agreement Pilot Project
TEDA	Tamil Nadu Energy Development Agency, India
TNEB	Tamil Nadu Electricity Board, India
TREN	Directorate General of the European Commission for Transport and Energy
TRIP	trade related aspects of intellectual property
TSECS	Tuvalu Solar Energy Cooperative Society
TW	terawatt (10^{12} W)
TWh	terawatt hour
UN	United Nations
UNCED	United Nations Conference on Environment and Development
UNCTAD	United Nations Conference on Trade and Development
UNDESA	United Nations Department of Economic and Social Affairs
UNDP	United Nations Development Programme
UNEP	United Nations Environment Programme
UNESCAP	United Nations Economic and Social Commission for Asia Pacific
UNESCO	United Nations Educational, Scientific and Cultural Organization
UNFCCC	United Nations Framework Convention on Climate Change
UNICEF	United Nations Children's Fund
UNIDO	United Nations Industrial Development Organization
US$	US dollars
USAID	United States Agency for International Development
USDOE	United States Department of Energy
USEPA	United States Environmental Protection Agency
VAT	value added tax
VBARD	Vietnam Bank for Agriculture and Rural Development
VDC	volts of direct current
VWU	Vietnam Women's Union
W	watt
W_e	watt of electricity
WBCSD	World Business Council for Sustainable Development
WEA	World Energy Assessment
WEC	World Energy Council

WEO	World Energy Outlook
Wh	watt hour
WHO	World Health Organization
WIPO	World Intellectual Property Organization
WMO	World Meteorological Organization
Wp	watt peak
WRI	World Resources Institute
WTO	World Trade Organization

Foreword

By Professor José Goldemberg

The case has often been made that fossil fuels are bad and renewable energy sources are good. Renewable energy sources are environmentally clean and locally available in most parts of the world, and some of them, such as fuelwood, are easily accessible to the poor. However, fossil fuels, although they are the main cause of many environmental problems, security of supply concerns and some social problems, today represent approximately 80% of the world's primary energy consumption, and this situation is unlikely to change significantly for some time.

Despite their attractiveness, renewable energy sources represent only 14% of all primary energy consumption and their use is distributed very unevenly. In the OECD they represent 10% of consumption and in developing countries 29%, most of which is used very inefficiently as non-commercial biomass. The contribution of 'new' renewables – wind, photovoltaics, solar, small hydropower, 'modern' biomass, geothermal and marine energy, excluding large hydro and non-commercial biomass – is just 2%. It is with these 'new' renewable energy sources that the author of this book is mainly concerned.

The question is frequently asked why 'new' renewables, if they are so attractive, are not used more widely. The main reason is cost; and it is reasonable to expect that, as has happened with many other technologies, the cost will fall as usage increases. This is why there is an important role for governments in introducing policies that will encourage the use of 'new' renewables, trying to make them more competitive. This applies for industrialized and developing countries alike. In the latter, however, there are additional barriers to the widespread use of renewables, notably access to technology, which do not apply in the former where the technologies were developed.

This book is a bold attempt to tackle the problems of transferring technology for renewable energies to developing countries. It addresses well both the problems facing attempts to transfer the technologies and ways of overcoming these barriers. The author uses her experience in Southeast Asia and concentrates her attention on two technologies: photovoltaics and biomass cogeneration. One might think that this is a narrow empirical base from which to draw general conclusions for developing countries, but in fact that is not so. The lessons learned in identifying barriers and how to remove them are of general application and will be very useful for training professionals interested in the subject, middle-level government officials, and experts in international lending organizations and NGOs. This is particularly important in the light of the progress achieved recently in Bonn and Marrakech on the ratification of the Kyoto Protocol and the operation of the Kyoto mechanisms, particularly the Clean Development Mechanism, which will benefit developing countries.

José Goldemberg
University of São Paulo
São Paulo, Brazil

Acknowledgments

The New Energy and Industrial Technology Development Organization of Japan (NEDO) funded the main research behind this book. NEDO undertakes various projects to promote new energy and energy conservation technologies in developing countries. The Royal Institute of International Affairs (RIIA) hosted the research. RIIA is an independent think tank that carries out research on important international issues. AEA Technology (AEAT) provided access to facilities during the final months of writing the text. AEAT is an energy and environmental consultancy based in Oxfordshire. The author was privileged to be involved in the G8 Renewable Energy Task Force, from which some information has also been drawn.

The author would like to thank the people, too numerous to mention individually, who spent time in meetings and provided information and comments which helped produce this book – you know who you are. Special thanks go to those who attended the study group: Katie Begg, Clive Caffall, Anthony Derrick, Gerald Foley, Michael Jefferson, Smail Khennas, Ritu Kumar, Gerald Leach, Justin Mundy, Stuart Parkinson, Norman Selley, Louise Simmonds, Garry Staunton and Christiaan Vrolijk. Thanks also go to those who peer-reviewed the book: Ron Alward, Tim Dixon, José Goldemberg, Chris Hazard, Debra Lew, Philip Mann, Eric Martinot, Anjali Shanker, Liam Slater, Julia Philpott and Neville Williams.

December 2001   *G.W.*

Introduction

Against a background of concern over poverty reduction, energy security and climate change, the transfer of renewable energy technology to developing countries is of interest to many organizations all over the world. Renewable energy has a key role to play not only in addressing emissions targets nationally and globally, but also in accessing local energy sources which can help facilitate sustainable development and meet international development targets. If the benefits it offers are to be maximized, renewable energy needs to be an integral part of national strategies for sustainable development (NSSD), poverty reduction strategy papers (PRSP) and any other development plans and targets. Renewable energy systems can provide electricity to remote rural populations that would not otherwise have access to it, providing energy services for homes, schools, and community and health centres. Renewable energy can also be a source of heat and mechanical power, providing a full range of services for households, communities, agriculture and small enterprises.

Harnessing indigenous sources of energy is an important element in the social and economic development of developing countries. Renewable energy technology can provide additional benefits such as increased employment, power for income-generating activities and a reduction in the use of fossil fuels. Moreover, a country which manufactures a renewable energy technology acquires the potential to export the technology to other developing countries where energy service requirements and environmental conditions are similar.

Some renewable energy technologies (e.g. wind turbines, small hydro technology and photovoltaic cells) have advanced considerably since the 1980s and are commercially deployed around the world. To date, industrialized countries have played a leading role in developing such technologies. To help developing countries leapfrog the heavy pollution and environmental degradation that went hand in hand with industrialization in the past, it is necessary to bring about the successful transfer to those countries of environmentally sound, appropriate, sustainable and commercially proven technologies. In pursuit of this general aim, it is important to understand the local situation in developing countries and meet the specific energy service demands with appropriate technology. One of the fundamental barriers which is often faced in transferring technology to developing countries is that the technology being transferred is not appropriate to the local context and demands, or is not adapted to the local environment. For example, electricity is certainly not the answer to all energy requirements: cooking is one of the largest elements of energy demand, and might more effectively be met by efficient cooking stoves using readily available local biomass or locally produced biogas. The right combination of energy sources and technologies needs to be identified for each situation. This might mean a combination of energy-efficient technologies powered by a range of renewables and fossil fuels.

The proposed creation under the Kyoto Protocol of the Clean Development Mechanism (CDM) has the potential to create greater opportunities for investment in renewable energy projects in developing countries, in exchange for emissions credits. At the time of writing, the CDM is yet to be agreed upon in detail, and the scale of the investment in renewable energy projects and its impact on renewable energy technology transfer will be determined by the eventual criteria for project selection and the ultimate structure and operation of the CDM. Nevertheless, in the build-up to agreement potential CDM projects are already being identified as targets for investment. With the CDM's emphasis on supporting sustainable development and reducing carbon dioxide emissions, renewable energy has the potential to feature strongly among the selected projects. However, investment alone will not guarantee an increase in the transfer of renewable energy technology, for there exists a range of barriers to the implementation of successful and sustainable renewable energy projects.

This book looks at the main barriers to the use of renewable energy in developing countries and options for overcoming them. The analysis is based on research undertaken for NEDO in Southeast Asia; in addition, a few other countries have been included where the author had experience and access to information. Material for this book has been collected using a combination of desktop research and visits to developing countries. When in developing countries, meetings were held with key actors in the renewable energy field, including where possible policy-makers, financiers and developers; technology manufacturers, installers and users; utilities, consultants and NGOs. The research is based on two particular renewable energy technologies (two were chosen to facilitate comparing lessons learned in different countries): stand-alone solar home systems in rural areas, and grid-connected biomass cogeneration. These technologies are not necessarily the most representative technologies for all developing countries, nor do they have the greatest energy generating potential; but they were chosen because they cover a broad range of potential barriers and issues. These two technologies highlight many of the general barriers faced by renewable energy technologies both on-grid and off-grid, but obviously do not cover the technology-specific issues faced by other particular technologies.

The barriers to photovoltaic (PV) solar home systems (SHS) related to the technology cost, and the ability and willingness of households to pay for the systems; the implementation and management of projects, e.g. local participation or the presence of an energy service company (ESCO), and the availability among people in local areas of the skills and knowledge necessary to install, operate and maintain the systems; the energy service requirements of communities and end users; the development of the potential market; access to remote areas for the supply of spare parts; and government policies on import duty and value added tax.

The barriers to biomass cogeneration related to the seasonal and geographical availability and price of biomass; competing uses for the resource base; access to the grid (power purchase agreements and liberalization of the electricity supply industry); government plans and targets for electricity generated from renewable resources; the perceived risk for financiers; use of inappropriate project appraisal methods; and lack of incentives for developers and entrepreneurs.

When looking for ways to overcome the barriers to renewable energy technology it is very difficult to pinpoint the specific conditions required for best practice, as in each situation there may be many local, national and international influences on the success or otherwise of a project. Some of these factors can have a direct and very obvious impact: for example, lack of access to credit. Other factors can have an indirect impact: for example, where a previous aid-funded project has provided electricity free of charge to nearby villages, people may be unwilling to pay for electricity even though they might value it highly and there may be a market for its use. However, some lessons regarding appropriate actions for overcoming barriers have been learned from both successful and unsuccessful projects. If such actions are considered when implementing new projects, the chances of project success can be increased, though not necessarily guaranteed. Factors specific to the particular region may impact favourably or unfavourably on a project in ways that are not predictable or obvious. Hindsight often provides the best vantage point from which to assess why projects have or have not worked. Improving the success of technology transfer will continue to be an ongoing process, as more projects are implemented, new problems are encountered and more lessons are learned.

Chapter 1 of this book looks at the potential role that renewable energy can play in sustainable development, security of energy supply and environmental protection, both locally and globally. Chapter 2 looks at the elements of technology transfer, from large-scale infrastructure development to the complementary transfer of skills and know-how. It outlines the key actors involved in technology transfer process, their roles, and the risks and rewards associated with their involvement. Chapter 3 looks at investment flows for technology transfer, the various sources of finance for renewables and innovative finance mechanisms, including the potential role of the CDM. Chapter 4 looks at the main barriers to technology transfer in developing countries and the various options for overcoming them. In particular, it summarizes barriers and options relating to stand-alone SHS and grid-connected biomass cogeneration gleaned from the case studies detailed in Annex 1. It also looks at options to facilitate the transfer of renewable energy technology in general. Chapter 5 highlights the common actions to facilitate technology transfer that can be taken by different actors and looks at the role for partnerships. It shows where certain actors can take the lead role or a supporting role, and which actions are most politically sensitive to implement. Some conclusions are drawn.

Annex 1 contains eight renewable energy case studies from developing countries and looks at the barriers and lessons learned in each. Annex 2 presents options for overcoming the barriers identified in the case studies.

Chapter 1

A Role for Renewables

1.1 Improving energy security

1.1.1 Trends in world energy use

Global demand for energy is rising. Recent predictions estimate that the 1997 total world primary energy demand of 8,610 million tonnes of oil equivalent (Mtoe) will have increased 30% by 2010 and 57% by 2020, reaching 13,529 Mtoe.[1] The current share of world primary energy supply provided by different fuels is predicted to change from 1997 to 2020 as follows: oil remaining around 40%; natural gas increasing from 22% to 26%; coal decreasing from 26% to 24%; nuclear decreasing from 7% to 5%; large hydro decreasing from 3% to 2%; and other renewables increasing from 2% to 3%.[2] Figures 1.1 and 1.2 show that although the shares of some fuels in world primary energy supply are likely to decrease or remain static, the supply of all fuels is predicted to increase in real terms. Thus it is predicted that there will be a continued reliance on fossil fuels for the next few decades at least.

The proportional reliance on fossil fuels is not in reality as great as shown in Figure 1.1, as these figures show only supply of commercial energy and thus leave out traditional biomass, which is a major source of energy used in developing countries. Traditional biomass is not bought or sold, so it does not enter into the commercial chain and therefore statistics on its use are patchy and unreliable. However, it has been estimated that traditional biomass accounts for around 34% of total primary energy

---

[1] *WEO* (2000), reference scenario data. 'Primary energy demand' is used interchangeably with 'total primary energy supply' and refers only to commercial energy use.

[2] The *WEO* (2000) definition of 'other renewables' in this context includes geothermal, solar, wind, tidal and wave energy. It also includes combustible renewables and waste (CRW) in OECD countries, but excludes it from developing countries, due to lack of accurate data. CRW comprises solid biomass and animal products, gas/liquids from biomass, industrial waste and municipal waste.

Figure 1.1: World primary energy supply, 1997 (Mtoe)

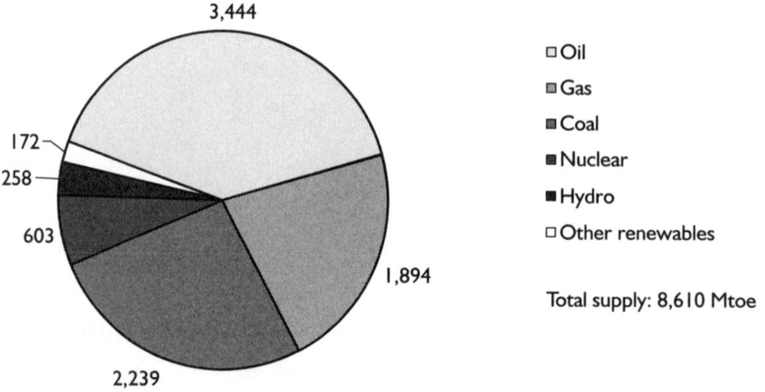

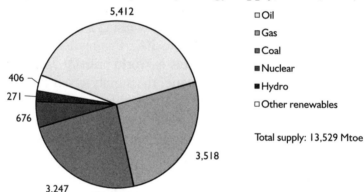

3,444
172
258
603
1,894
2,239
Oil
Gas
Coal
Nuclear
Hydro
Other renewables
Total supply: 8,610 Mtoe

Source: Based on figures from *WEO* (2000).

Figure 1.2: World primary energy supply, 2020 (Mtoe)

5,412
406
271
676
3,518
3,247
Oil
Gas
Coal
Nuclear
Hydro
Other renewables
Total supply: 13,529 Mtoe

Source: Based on figures from *WEO* (2000).

consumption in developing countries, represents more than 90% of total energy demand in some of the least developed countries and is predicted to continue to play an important role over the next few decades.[3]

The projected growth in energy supply and relative percentages of different energy sources shown in Figure 1.2 are based on the IEA reference scenario where growth in demand is based on continued present

[3] FAO (1999).

Box 1.1: Energy sources and terms used

Energy sources

There are many different sources of energy, including:

- *fossil fuels:* coal, natural gas and oil (including derivatives such as kerosene, gasoline and diesel);
- *nuclear;*
- *renewable energy:* large and small hydro;[a] solar; wind; geothermal; energy from oceans (wave, tidal and thermal); biomass (wood, dung, leaves, crop residues and animal and human waste);
- *municipal and industrial waste* (used for combustion or production of gas).

Terms used in this book

- *Conventional fuels* include fossil fuels, large hydro and nuclear.
- *Traditional biomass* refers to the traditional, often inefficient, use of biomass resources, for example in open fires, three-stone fires or wood and charcoal stoves.
- *Modern biomass* is the cleaner, more efficient use of biomass resources, for example in efficient cooking stoves, for electricity generation (e.g. combined heat and power), and in liquid biofuels and biogas.
- *New renewables* exclude large hydro and traditional biomass.
- *Combustible renewables* include solid biomass and animal products, and gas and liquids from biomass.
- *Combustible renewables and waste* (CRW) includes both industrial waste and municipal waste.
- *Commercial energy* means energy that is bought and sold in the marketplace e.g. electricity, fossil fuels and charcoal.
- *Non-commercial energy* means fuels that are gathered free of charge e.g. wood, leaves, agricultural residues and dung.

[a] Small hydro is defined in various ways around the world, from 1 MW up to 50 MW in some cases. This book uses a definition of up to 15 MW for small hydro.

---

trends. The scenario aims to take account of a range of new policies in OECD countries designed to combat climate destabilization. The policies included were all enacted or announced by mid-2000, though may not have been fully implemented as yet. A number of other organizations have looked at different scenarios and made projections of energy supply into the future, and this IEA scenario is felt by some to be rather conservative with respect to the growth of renewable energy. This is emphasized even more by the omission of traditional biomass from the figures. However, for the purposes of this book the IEA figures will be used as a basis for some broad analysis of the likely impact of the CDM on renewable energy uptake (see Chapter 3).

The two most significant factors influencing growth of world energy demand are economic growth and population growth. It is expected that nearly 70% of the growth in primary energy demand between 2000 and 2020 will come from developing countries and economies in transition (see Figures 1.3 and 1.4). This is predicted to be due to high rates of population increase and urbanization, rapid industrial expansion, economic growth and a continued move to commercial fuels from non-commercial fuels.

Figures 1.3 and 1.4 show the estimated change in world primary energy supply by region from 1997 to 2020. They show a predicted increase in demand in developing countries from 34% in 1997 to 45% in 2020, China being the developing country with the greatest increase. In 1997 OECD countries accounted for just over 1 billion people (around one-

Figure 1.3: World primary energy demand by region, 1997 (%)

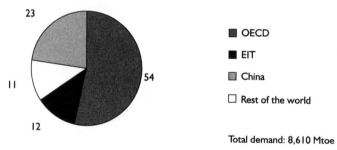

Source: WEO (2000).

Figure 1.4: World primary energy demand by region, 2020 (%)

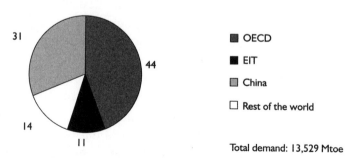

Source: WEO (2000).

Table 1.1: World total final energy consumption (Mtoe)

	1997	2010	2020	Average annual growth rate, 1997–2020 (%)
Total final consumption	5,808	7,525	9,117	2.0
Coal	635	693	757	0.8
Oil	2,823	3,708	4,493	2.0
Gas	1,044	1,338	1,606	1.9
Electricity	987	1,423	1,846	2.8
Heat	232	244	273	0.7
Renewables	87	118	142	2.2

Source: Data taken from *WEO* (2000).

sixth of the world population) and a staggering 54% of world energy demand, whereas in developing countries and economies in transition around five billion people consumed only 46% of world energy supply. This made the average energy consumption per capita in OECD countries around six times higher than in developing countries and EITs. The comparison is even more stark if one looks only at the least developed countries. These regional statistics hide the fact that on an annual basis the 2 billion poorest people, whose income is US$1,000 or less, use barely 0.2 toe of energy per capita, whereas the billion richest people, whose income is US$22,000 or more, use approximately 5 toe per capita – around 25 times as much.[4]

Projected demand for electricity will grow more rapidly than that for any other form of commercial energy, at an average rate of increase of 2.8% per year from 1997 to 2020. Table 1.1 shows that demand for electricity is expected to double from 1997 to 2020. It is estimated that, today, around 1.7 billion people (nearly 30% of the world's population) in developing countries are without access to electricity.[5] The population in developing countries is increasing rapidly and although electrification is taking place in many of these countries, it is not keeping pace with population growth. But in countries with strict birth control measures and laws (e.g. China) electrification rates are keeping up with and even exceeding population growth.[6] Around 3,000 GW of new generating capacity will

---

[4] G8 RETF (2001), p. 13.
[5] *WEA* (2000), p. 44.
[6] World Bank (1996), p. 2.

need to be built between 1997 and 2020, more than half of it in developing countries.[7]

Population and energy demand increases in developing countries highlight the urgent need for sustainable, affordable and environmentally sound energy systems. The choice of energy technologies (in particular large power-generating technologies) among those available to developing countries in the next few years is extremely important if they are to develop sustainable energy supply networks in the future, due to technology lock-in effects. That is, it is costly to invest in new generation technology, so once the country has selected and invested in a technology it will continue to use it for many years, until its useful life is over. For their part, industrialized countries need to continue developing and deploying renewable energy technology in their own countries to help reduce technology costs, mitigate environmental impacts and develop the range of technologies needed for sustainable energy supply.

1.1.2   *Current deployment of renewable energy and future prospects*

Modern biomass has an important role to play in providing cleaner, more efficient energy services for cooking and space heating in developing countries. Another promising application for new renewable technologies, particularly in the short to medium term, is electricity generation. Table 1.2 shows electricity generation from non-hydro renewable energy in 1997, broken down by region and energy source. These figures indicate that total capacity in 1997 was around 48 GW. Approximately 77% of this capacity is deployed in the OECD, with Asia also taking a significant share (10%). Of the non-hydro renewables, combustible renewables and waste (CRW) have the highest installed capacity, accounting for 65% of the total, and wind and geothermal have shares of 17% and 16% respectively. As these figures exclude hydro, it is worth noting that an energy technology study funded by the European Commission estimated that there was around 28 GW of small hydro capacity worldwide in 1995.[8]

---

[7] *WEO* (2000), Table 1.4.
[8] CEC, DGXVII (1997).

Table 1.3 shows the amount of additional non-hydro renewable energy electricity generating capacity forecast to be built over the period 1997–2010. It shows that a total of around 43 GW of additional capacity is predicted to be built, almost doubling the total installed capacity. The table shows that most of this new capacity is likely to be built in

Table 1.2: Non-hydro new renewable electricity generating capacity, 1997 (GW)

Region	Geothermal	Wind	CRW	Solar/tide/ other	Total
OECD Europe	0.6	4.5	6.5	0.5	12.1
OECD North America	2.9	1.7	13.5	0.4	18.5
OECD Pacific	0.9	0.0	5.9	0.0	6.8
EIT	0.0	0.0	1.5	0.0	1.5
Africa	0.1	0.1	0.2	0.0	0.4
China	0.1	0.4	0.0	0.0	0.5
East Asia	2.1	0.0	0.7	0.3	3.1
Latin America	1.1	0.3	2.8	0.0	4.2
Middle East	0.0	0.0	0.0	0.0	0.0
South Asia	0.0	1.0	0.2	0.0	1.2
Total	7.8	8.0	31.3	1.2	48.3

Source: Data taken from *WEO* (2000).

Table 1.3: Projected increase in non-hydro new renewable electricity generating capacity, 1997–2010 (GW)

Region	Geothermal	Wind	CRW	Solar/tide/ other	Total
OECD Europe	0.4	16.8	5.3	1.1	23.6
OECD North America	0.1	3.9	2.1	0.4	6.5
OECD Pacific	1.2	1.0	0.7	0.3	3.2
EIT	0.1	0.2	0.1	0.0	0.4
Africa	0.2	0.3	0.2	0.0	0.7
China	0.2	1.9	0.1	0.1	2.3
East Asia	1.2	0.0	0.1	0.1	1.4
Latin America	0.3	0.3	1.2	0.0	1.8
Middle East	0.0	0.1	0.0	0.0	0.1
South Asia	0.0	1.8	0.8	0.2	2.8
Total	3.7	26.3	10.6	2.2	42.8

Source: Data taken from *WEO* (2000).

Table 1.4: Global new generating capacity, 1997–2020 (GW)

Region	New generating capacity	
	1997–2010	*2010–2020*
OECD Europe	229	248
OECD North America	195	201
OECD Pacific	64	75
OECD total	*488*	*523*
EIT	116	222
Africa	47	57
China	253	266
East Asia	122	156
Latin America	139	143
Middle East	54	77
South Asia	115	135
Developing countries total	*730*	*834*
World total	*1,335*	*1,579*

Source: WEO (2000).

those regions where significant capacity already exists, such as the OECD and Asia, and that the largest increase in non-hydro capacity is expected from wind.

Table 1.4 presents the projected increase in all global electricity generating capacity between 1997 and 2020. It shows that more than 1,300 GW of additional electricity generating capacity (from all sources of energy, including renewables, fossil fuels and nuclear) will be required worldwide by 2010, and approximately 1,600 GW of further electricity generation capacity by 2020. More than half of this new capacity will be in developing countries.

Around one-fifth of the new generating capacity required between 1997 and 2020 is predicted to replace existing installations; the remainder will be to meet new demand. Over the period from 1997 to 2010, China is likely to need around 250 GW of new capacity and the rest of Asia and Latin America over 375 GW. Although the projected increase in non-hydro new renewable energy generating capacity from 1997 to 2010 is significant in terms of existing such capacity (i.e. it almost doubles), it is equivalent to less than 4% of the total projected increase in generating capacity over this period. Unless the growth rate of electricity and other forms of energy from renewables is at least 100% per year, renewables will not be a significant contributor by the year 2020. *WEO* 2000 estimates indicate that the investment required for new generating capacity in developing countries is US$870 billion from 1997 to 2010 and US$839 billion from 2010 to 2020, totalling US$1.7 trillion. This sum excludes investment needed for transmission and distribution lines, which will be high in developing countries where existing geographical coverage is low. Taking this into account could easily double the total investment requirements in developing countries. Additional investment will be needed to realize a larger percentage of the potential for renewables. Many developing countries are beginning to recognize the role of private investment as an option to help expand their power sectors, thus assisting development. Chapter 3 looks at investment in renewable energy and the potential role of the Clean Development Mechanism in channelling funds towards renewable energy technology transfer.

1.1.3   The need for indigenous local energy sources

Dependence on oil imports in many developing countries makes them susceptible to the impacts of oil price shocks. As we saw in 1999–2001, oil price spikes are not a the thing of the past but a very real

Table 1.5: Projected net oil imports and exports (million barrels per day)

Region	1997	2010	2020
OECD North America	9.0	12.6	15.2
OECD Europe	7.4	10.8	13.3
OECD Pacific	5.7	6.4	6.6
EIT	−2.8	−4.5	−4.9
Africa	−6.1	−9.4	−9.5
China	0.9	4.6	8.5
Other Asia	4.9	10.8	16.7
Latin America	−4.1	−5.4	−4.6
Middle East	−17.0	−26.6	−41.3

Source: WEO (2000). Negative numbers indicate net exports.

threat that will continue as long as demand for oil is high. Table 1.5 shows which regions of the world are net oil importers and exporters. Latin America and Africa as a whole are net oil exporters; however, there are many countries within these regions that are net oil importers. The table shows that between 1997 and 2020 the volume of oil imports is predicted to rise from 0.9 to 8.5 million barrels per day in China and from 4.9 to 16.7 million barrels per day in other Asian countries.

Oil import dependence is growing in proportional as well as in real terms. Table 1.6 shows dependence on imported oil, as a percentage of all oil used, for different regions. The predictions show that from 1997 to 2020 China's dependence on oil imports is likely to increase from 22% to 77%, India's from 57% to 92%, East Asia's from 54% to 81% and that of the rest of South Asia from 87% to 96%. It is clear that with such high dependence on imported oil predicted for the future there will be a growing concern in developing countries over security of energy supply. To increase security of supply a greater mix of fuels is needed, so that both dependence and the associated risks are spread. Ironically, poorer and more isolated countries already pay more for imports of petroleum products as they are not in a strong position to negotiate the price down. Increased dependence on oil imports and potential future price fluctuations make these countries more vulnerable to macroeconomic impacts, reducing the availability of funds for poverty-reducing initiatives, for example in health and education.

As well as increasing the mix of fuels used, it is important that fuels be used more efficiently. There are opportunities to improve economic

Table 1.6: Oil import dependence (%)

Region	1997	2010	2020
North America	44.6	52.4	58.0
Europe	52.5	67.2	79.0
Pacific	88.8	91.5	92.4
All OECD	54.3	63.3	70.0
China	22.3	61.0	76.9
India	57.4	85.2	91.6
Rest of South Asia	87.2	95.1	96.1
East Asia	53.7	70.5	80.7

Source: WEO (2000).

growth and the standard of living of the world's population if energy demand increases more slowly than the historical trend. Energy-saving measures and greater energy efficiency on both the demand and the supply side can help slow down the upward trend in energy demand. Most industrialized countries have fiscal measures and standards in place to try to improve the efficiency with which energy is provided and used. Households and businesses in developing countries are not likely to invest in energy-efficient technology (cooking stoves, boilers etc.) that save energy unless they are made aware of the benefits to them physically (health, reduced time and drudgery), economically (overall saving on cost), or environmentally (reduced local or indoor air pollution). But even then they may not have access to capital to invest in more efficient technology. However, some energy efficiency measures involve little or no cost: improved management of industrial energy use, for example.

Threats to security of supply and the projected levels of new generating capacity that will be needed highlight the urgency of starting to address some of the barriers to renewable energy technology transfer and to introduce policies and measures to encourage renewable energy use and energy efficiency. Indigenous sources of renewable energy have the potential to play a key role in securing supply. Action needs to be taken urgently to expand the deployment of renewable energy technology, as it will take time to accumulate the number of systems required, as well as the supporting infrastructure and local skills needed to install and maintain these systems. This is partly because renewable energy systems are relatively small-scale and therefore large numbers will need to be installed before they have a significant impact on projected oil import dependence patterns.

1.1.4   A niche for renewable energy

Electricity and fossil fuels in developing countries are often subsidized by government, thus making it harder for renewable energy systems to compete economically. Renewable energy systems are therefore most likely to be installed in niche markets in areas where it is difficult and expensive to transport fossil fuels and which are not grid-connected.

Grid extension is being carried out in many developing countries where possible, but its being technically possible does not mean it is cost-effective. Urban areas are densely populated and thus one electricity line can supply a large number of people in a small geographical area. The opposite is true for rural areas, where populations and hence demand for electricity are widely spread out. One electricity line will typically serve many fewer people in a rural area, and therefore the investment cost per head of population is much greater. In state-owned electricity supply industries, the true cost of grid extension is often hidden among other costs and subsidies. With grid electricity in rural areas it is difficult to match the level of supply with the level of demand, as electricity lines will typically provide much more power than households initially require, due to the absence of household electrical appliances. As households can afford to buy more appliances, the demand will grow to match supply. In such cases the electricity can be described as supply-driven. There is generally a better match in supply and demand if electricity is demand-driven, or market-led. An example of this might be an entrepreneur setting up a battery-charging station where he knows there will be sufficient demand for it to be economic. His business may start on a small scale, but can expand as demand rises.

The niche markets for renewable energy systems in developing countries are most likely to be located off-grid in rural areas. There are several energy demand sectors in rural areas, including household, farming, industry, offices and shops, transport and community services (schools, health centres, water pumping, street lighting, etc.). As rural areas can be remote and difficult to reach, it is important to have appropriate supporting infrastructure for the renewable energy systems. This support should include training and education to develop local technical knowledge and expertise to operate and maintain the new renewable energy system; access to spare parts; and access to credit or the presence of an energy service company (ESCO) with leasing arrangements.

Various forms of energy (light, heat, mechanical power, electricity) will be needed in order to fulfil energy requirements, for example for cooking, space heating, crop processing, lighting and communications. Renewables are not necessarily the best option to provide all the required energy services and should be considered in combination with other energy sources and technologies to create a package of energy services that provides all the required needs in the most convenient, economic and environmentally sound way. In South Africa this approach is described as 'energization'.[9]

1.2 Powering sustainable development

1.2.1 Energy services

There is a growing gap between rich and poor, industrialized and developing countries. Energy has a key role to play in bridging that gap. According to United Nations statistics, the world population officially reached 6 billion people on 12 October 1999. Of these, approximately 2 billion people are estimated to be without access to adequate, affordable, clean sources of energy and are reliant on traditional biomass for cooking and space heating.[10] The majority of these people live in rural areas of developing countries, but there is also a growing problem in

Figure 1.5: Energy, a derived demand

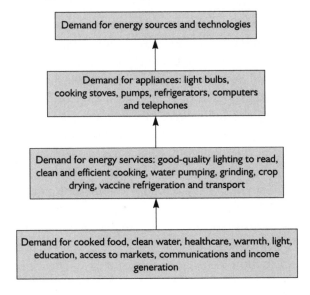

---

[9] Personal communication with Chris Hazard, PN Energy, South Africa.
[10] World Bank (1996).

urban and peri-urban areas. World population is expected to grow from 6 billion to 7.4 billion by 2020, the majority of the expansion predicted to take place in urban areas of developing countries.[11] In 1997 around 77% of world population lived in developing countries; it is predicted that by 2020 this percentage will reach 81%.[12]

Energy use is a derived demand (see Figure 1.5): that is, most people do not consciously desire a particular fuel or form of energy, but desire the service it provides (e.g. cooked food, communications, transport, light to read). People all over the world need light to see at night, heat to cook and keep warm, mechanical power to pump, grind and provide transportation and electricity to power telecommunications, computers and other electrical appliances. Energy used for cooking and space heating is the first priority in developing countries, followed by electricity for lighting, water pumping, radio, communications and other appliances.

In stark contrast to people living in industrialized countries, who without thinking just flick a switch for light or turn a knob for heat, those living in rural areas of developing countries spend a lot of time and effort accessing energy and are more conscious than most of the fuel that provides desired services. People in developing countries, particularly women and children in rural areas, regularly spend several hours every day collecting wood, leaves, dung and other forms of biomass such as straw, which they burn to cook and keep warm. For example, in the hill areas of Nepal, where supplies of wood are fairly good, women need to spend over an hour a day collecting fuel; and in areas where wood is more scarce, this task can take up to 2.5 hours a day.[13] Biomass still accounts for around 80% of all household fuel consumption in developing countries. Women are the main collectors and users of biomass energy, often having to carry heavy loads over long distances. They also have to carry water over long distances. Carrying heavy loads on their heads can lead to neck and back injury, pain and headaches. It also reduces time available to women to care for infants and to carry out income-generating activities that can make an important contribution to household income.

---

[11] *WEO* (2000), p. 21.
[12] Ibid.
[13] World Bank (1996), p. 1.

In developing countries there is a direct link between poverty and dependence on biomass for energy. The poorest families use leaves, animal dung and wood as fuel, which can often be collected free of charge. As families become wealthier they can afford to use other forms of energy if they are available (kerosene, gas or electricity). Economic growth or access to credit is needed to enable the uptake of more modern energy sources and technologies. If conditions are right (i.e. there are other natural resources available and access to markets), the services provided by the new energy source can be used for income-generating activities (e.g. baking food to sell, food processing, making handicrafts after dark with the aid of electric lighting). In addition, installing, maintaining or manufacturing the new energy systems and technologies can create jobs. For example, it is reported that around 43 jobs have recently been created in a group of renewable energy projects in the Maphephetheni community in KwaZulu Natal in South Africa.[14] The projects included providing electricity for computers in schools, electricity and hot water for health clinics, pay cellphones, solar cooking, crop drying, biogas from installation of latrines, water pumping, irrigation and energy services for tourism. With international aid programmes being driven to a greater extent by the imperative of alleviating poverty,

Table 1.7: Renewable energy sources and forms of energy

Energy source	Form of energy			
	Light	Heat	Mechanical power[a]	Electricity
Solar	Sunlight	Passive solar Solar water heating Solar drying		Photovoltaics
Wind			Wind pump	Wind turbine
Hydro			Water mill	Water turbine
Biomass[b]	Flame	Combustion[c]		Steam turbine[c]

a Electric pumps can be powered by any of the energy resources in this table.
b Liquid or gaseous biofuels can be produced for use in internal combustion engines for electricity generation or mechanical power.
c Combined heat and power (CHP) is an efficient way to generate heat and electricity.

---

14 Cawood (2001).

it is important to understand more fully the role that renewable energy systems can play in addressing priority development objectives in, for example, the health and education sectors and in developing rural enterprises, empowering women, enabling income-generating activities and creating employment.

The renewable energy resources most widespread and readily available in developing countries are solar, wind, hydro and biomass (although geothermal resources have been harnessed in some countries for many years and accounted for 16% of total world electricity production from new renewable energy systems in 1997[15]). These resources can be harnessed in a number of ways to produce light, heat, mechanical power or electricity. Table 1.7 shows the main forms of energy available from these four energy sources.

Solar   The heating properties of solar radiation can be harnessed in various ways, including solar water heaters and solar dryers (in agriculture). Alternatively, solar photovoltaic (PV) cells can be used to generate electricity from the sun. The PV cells can be connected directly to a pump for irrigation or drinking water, or they can be connected to a storage battery for use in houses, schools, shops or medical centres, providing electricity for lighting, radio, television, video, or refrigeration or other appliances.

Wind   Wind can be harnessed for pumping irrigation or drinking water, or for electricity generation via a wind turbine. As with PV systems, electricity can be used directly or to charge a battery for later use.

Hydro   Water can be used for mechanical power (water mills for grinding or pumping), or for electricity generation via a turbine. Large-scale hydropower is not now considered to be very environmentally friendly because of the displacement of populations, loss of biodiversity and disruption of aquatic life involved. However, small-scale hydro up to 15 MW (including mini, micro and pico hydro) of the run-of-river or small dam construction type is much more environmentally benign.

---

[15] See Table 1.2.

Biomass   Traditional uses of biomass, for example using open fires for cooking, are very inefficient. There are many different biomass sources available that can be used in more efficient or 'modern' ways to provide energy services in developing countries.[16] These include efficient stoves for cooking, or electricity from combined heat and power plant using agricultural residues (from palm oil, sugar or rice processing), sawmill residues or dedicated crops (wood from sustainable forestry plantations). In addition, animal slurry (and human night soil) can be used for biogas production via anaerobic digestion. Fermentation, liquefaction and gasification are other methods of generating liquid and gaseous fuels from biomass resources that are at various stages of development and deployment. Biomass can be a sustainable source of energy if used efficiently and managed well. If, however, resources are overused and mismanaged, serious harm can ensue: for example, in the case of wood fuel, deforestation can take place and there can be permanent damage to the land area. In extreme cases excessive collection of wood fuel can be as damaging as logging or slash-and-burn clearance of land for agriculture, and can lead to desertification. If dung and crop residues are overused as fuel, there is not enough left to use as fertilizer, and crops and soil conditions will suffer. Therefore, good management and efficient use of biomass resources are important for sustainable development.

In order for renewable energy systems to be sustainable they must be economically viable and environmentally sound, and provide a service that is needed. The energy demand of a community is directly linked to its socio-economic conditions, the local environment and its current technical status (e.g. level of energy efficiency and access to technology). The wants and needs of a community are often different from their ability to pay for services. It is very important when planning new energy systems that they be chosen and designed with the market or demand for energy service in mind. For new renewable energy systems to be successful, there must be a market for the service, and it must be possible to provide it at a price which is affordable to the users.

---

[16] The term 'traditional biomass' refers to traditional inefficient methods of using biomass for energy. The term 'modern biomass' refers to more efficient and cleaner ways of using biomass for energy, using modern technology.

In order to match the demand in energy service with the appropriate energy technology it is very important to carry out an energy survey of the required services and to include local participation in this process. It is also very important to integrate energy into development planning, i.e. to include energy issues in national development plans. For example, when planning new health centres or schools, it is important to look at their energy requirements and plan sustainable energy systems to be installed at the same time as the building is developed, including energy-efficient building design. The selection of the particular technical option should be based on three main criteria:

- service provided (availability and applicability);
- affordability (initial investment and running costs);
- sustainability (environmental, technical and institutional).

Energy is not required just to provide energy services; it is fundamental to increasing productivity and generating revenue streams and jobs. Energy is an underlying resource that is required to assist poverty reduction.

1.2.2   Energy and development

Sustainable development and the role of energy in the development process are vital issues that have been gaining more attention and concern over the last few decades (see Box 1.2). The core question is: How can we continue to meet our needs and develop on a global basis while at the same time protecting the environment and natural resources of the planet for current and future generations?

There is no standard sustainable development agenda for developing countries; the precise form of each will differ depending on country context, requirements and priorities. Within the framework of Agenda 21 (the programme for action aimed at achieving sustainable development in the twenty-first century), each country can define its programme differently. The basis for Agenda 21 was agreed during the 'Earth Summit' at Rio de Janeiro in 1992 and signed by 179 heads of state and government. Among other things, it calls for all countries to set national strategies for sustainable development (NSSD), recognizing that one sustainability

Box 1.2: Growing interest in sustainable development and the role of energy

The list below highlights a few notable publications and events from a growing number over recent years that have helped to build awareness and understanding of the need for sustainable development and the importance of environmentally sound energy technologies and fuels.

1979 – First World Climate Conference and establishment of the World Climate Research Centre
1987 – World Commission on Environment and Development, *Our Common Future* (the Brundtland report)
1988 – The Intergovernmental Panel on Climate Change set up under UNEP and WMO
1989 – Pearce et al. *Blueprint for a Green Economy*
1992 – Conference on Environment and Development, the first 'Earth Summit' on sustainable development (Rio de Janeiro). Signing of the United Nations Framework Convention on Climate Change (UNFCCC) and Agenda 21
1994 – UNFCCC established
1994 – Conference on Population and Development
1994 – Global Conference on Small Island Developing States (SIDS)
1995 – World Summit for Social Development
1995 – Beijing Fourth World Conference on Women and Development
1995 – UNDP report, *Energy as an Instrument for Socio-Economic Development*
1996 – World Bank report, *Rural Energy and Development: Improving Energy Supplies for Two Billion People*
1997 – UNDP/IEI/SEI report, *Energy after Rio: Prospects and Challenges*
1997 – Kyoto Protocol signed at the third Conference of the Parties to the UNFCCC
1999 – UNDP and European Commission report, *Energy as a Tool for Sustainable Development for ACP Countries*
1999 – FAO/WEC report, *The Challenge of Rural Energy Poverty in Developing Countries*
2000 – UNDP/UNDESA/WEC report, *World Energy Assessment: Energy and the Challenge of Sustainability*
2000 – WEC report, *Energy for Tomorrow's World – Acting Now!*
2000 – ESMAP energy and development report, *Energy Services for the World's Poor*
2000 – International development targets (IDTs) agreed by G8 leaders and adopted by development agencies.
2000 – UNDP initiative for sustainable energy (UNISE) report, *Sustainable Energy Strategies: Materials for Decision Makers*
2000 – UN–ESCAP meeting on 'Energy for Sustainable Development' (Bali)
2000 – First Global Forum on Sustainable Energy, 'Rural Energy: Priorities for Action' (Austria)
2000 – FAO report, *The Energy and Agriculture Nexus*
2001 – Commission on Sustainable Development meeting, CSD9 (New York), focused on energy issues, with a build-up to Rio + 10 to be held in South Africa in 2002
2001 – G8 Renewable Energy Task Force Report

agenda does not fit all countries. These NSSDs are scheduled to be completed by 2002 and implemented by 2005.

The big challenge is how to achieve sustainable development for the world's population, around 80% of whom live in developing countries, without placing huge burdens on the people least able to bear them. Developing countries are encouraged by the World Bank to formulate poverty reduction strategy papers (PRSPs) in order to help channel funding from multilateral and bilateral agencies with greater effect in achieving their development and environment goals. The core principles of PRSPs are that they should be country-driven, results-oriented, comprehensive in scope, partnership-oriented, long-term in perspective and participatory. As with NSSDs, there is no 'blueprint' for PRSPs. However, there are three key steps: understanding the features of poverty and the factors that determine it; choosing public actions which have the greatest impact on poverty; and identifying indicators of progress and monitoring them in a participatory manner. After July 2002 all World Bank International Development Association (IDA) country assistance strategies must be based on PRSPs.

1.2.3 *Achieving international development targets*

Bilateral and multilateral development agencies have adopted a common set of international development targets (IDTs; see Box 1.3). These targets aim worldwide to reduce poverty, improve access to education, empower women through equality, reduce infant and maternal mortality rates, slow population growth through contraception and implement national strategies for sustainable development. More recently, a set of millennium development goals (MDGs), which are very similar to the IDTs, has been adopted by some development agencies.

Poverty is often conceptualized and measured in terms of the proportion of people who do not achieve specific levels of health, education or body weight. It can also be measured in terms of the number of people without the monetary resources that would enable them to consume a defined bundle of goods and services for basic survival. The energy dimension of poverty can be defined as the absence of sufficient choice in accessing adequate, affordable, reliable, high-quality, safe and

Figure 1.6: Energy and social issues

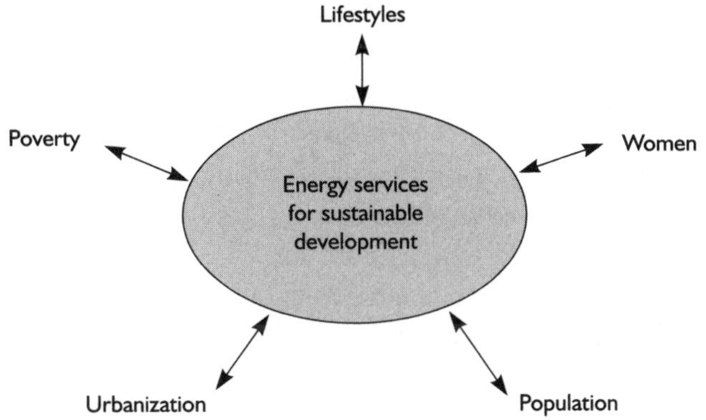

Source: Adapted from *WEA* (2000), ch. 2, p. 43.

Box 1.3: International development targets

- A reduction by one-half in the proportion of people living in extreme *poverty* by the year 2015.
- *Universal primary education* in all countries by 2015.
- Demonstrated progress towards *gender equality* and the empowerment of women by eliminating gender disparity in primary and secondary education by 2005.
- A reduction by two-thirds in the *mortality* rates for infants and children under age five and a reduction by three-quarters in maternal mortality, both by 2015.
- Access through the *primary healthcare* system to reproductive health services for all individuals of appropriate ages as soon as possible and no later than 2015.
- The implementation of national strategies for *sustainable development* by 2005, so as to reverse current trends in the loss of environmental resources at both global and national levels by 2015.

---

environmentally benign energy services to support human and economic development.[17] Estimates suggest that 2 billion people are without clean, safe cooking fuels and depend on traditional biomass sources.[18]

Increased access to modern energy services will not, in itself, result in economic or social development, but it is necessary to assist development. Figure 1.6 shows that energy services interact with social

---

[17] *WEA* (2000), p. 44.
[18] World Bank (1996), p. 15.

issues and can result in changes in lifestyle that may contribute to empowerment and equality for women, poverty reduction, reduced urban migration and reduced rates of population increase.

Table 1.8 illustrates how energy can play an important role in achieving the IDTs, both directly and indirectly. Renewable energy has an important role to play as it can be accessed locally in rural areas, making it the most convenient and economic option in many cases.

Poverty alleviation   Energy is fundamental for economic growth. Access to reliable energy services can directly contribute to enterprise development and income-generating activities if other raw materials and markets are accessible. Lighting to extend the working day and machinery both facilitate increased productivity. Productive activities after dark such as sewing, handicrafts and longer opening hours for stores, plus access to other technologies such as telecommunications, can increase local incomes, improve quality of life and help reduce migration of rural people into towns and cities. Local jobs can be created in energy service provision (e.g. installing, maintaining and operating the technology and in growing dedicated fuel crops), thus also contributing to poverty alleviation. Increased use of renewable energy in combination with energy efficiency can help reduce both dependence on fossil fuels and imports, contributing to a more stable macroeconomic environment.

If families have access to more efficient cooking stoves, biogas or electricity, they do not need to spend so much time collecting fuelwood and other biomass resources. They have the option of spending more time doing productive activities such as farming or making handicrafts and food to sell. The same can be said for water collection. Women and children often spend a significant amount of time each day fetching water. This burden can be reduced by the use of renewable energy systems for pumping drinking water.

Education and equality for women   Modern energy has a fundamental impact on the feasibility of children and adults (particularly girls and women) attending school, as it alleviates the time spent on survival activities such as gathering fuelwood, fetching water, cooking over inefficient fires and food processing. Good-quality lighting enables evening classes and study at night. Electricity in schools makes possible the use of overhead projectors, computers, printers, photocopiers, TV, video, radio and cassette, all of which directly contribute to the quality and standard of educational material available to pupils. Electricity also allows access to the internet for educational material and communications. Access to education helps empower women.

Apart from freeing up women's time for education, modern energy services are also fundamental to improving the condition of women by reducing the drudgery of survival activities (such as the need to carry heavy loads of fuelwood and water, which can cause neck and back injury and headaches). Having the choice of whether and when to have children also helps empower women, and energy is fundamental to manufacturing contraceptives and transporting them to health clinics and shops. Reliable energy services offer scope for women's enterprises (for example, smoking fish, weaving mats, baking bread), and street lighting improves women's safety.

Health Open fires give off smoke and, if inside, cause indoor air pollution. Smoke from biomass fires contains hazardous chemicals that vastly exceed the WHO recommended levels for safety. As many as 1 billion people, mostly women and children, are regularly exposed to levels of indoor air pollution exceeding WHO guidelines by up to 100 times.[19] Studies have shown that women cooking indoors over open fires suffer disproportionately from respiratory illnesses and eye infections. Small children exposed to indoor air pollution also suffer from such illnesses. Pregnant women exposed to high levels of indoor air pollution are more likely to have a miscarriage or stillbirth. Exposure to smoke can cause premature deaths in women and children. Acute respiratory infections are the fourth most prevalent group of diseases in sub-Saharan Africa.[20] In India, the pollution from household solid fuel use causes an estimated 500,000 premature deaths a year in women and children under the age of five.[21] Exposure to smoke and soot has

---

[19] WHO (2000).
[20] WHO (1999).
[21] *WEA* (2000), p. 69.

Table 1.8: Energy and the international development targets

Target	Fundamental	Importance of energy to achieving the target	
		Contributes directly	Contributes indirectly
A reduction by one-half in the proportion of people living in extreme *poverty* by the year 2015	• Modern energy supplies are necessary for economic growth • Clean, efficient fuels can reduce the amount of time spent on collecting biomass fuels, allowing more time for income-generating activities	• Access to reliable energy services enables enterprise development • Lighting permits income generation beyond daylight hours • Increased productivity from being able to use machinery	• Employment creation in local energy service provision and maintenance, fuel crops, etc.
Universal *primary education* in all countries by 2015	• Availability of modern energy services frees children's time from helping with survival tasks (gathering firewood, fetching water)	• Good-quality lighting permits home study • Electricity in schools helps retain teachers and creates a more child-friendly environment, reducing dropout rates.	• Electricity enables access to educational media in schools and at home via computers, printers, overhead projectors, photocopiers, TV, video, radio/cassette and the internet
Demonstrated progress towards *gender equality* and the empowerment of women by eliminating gender disparity in primary and secondary education by 2005	• Availability of modern energy services frees women's time from survival activities (fuel gathering, cooking inefficiently, fetching water, crop processing by hand, manual farming work)	• Good-quality lighting permits home study • Electricity in schools allows evening classes and helps retain teachers	• Electricity enables access to educational media • Reliable energy services offer scope for women's enterprises • Street lighting improves women's safety
A reduction by two-thirds in the *mortality* rates for infants and children under age five and a reduction by three-fourths in maternal mortality, all by 2015	• Indoor air pollution from traditional fuels causes significant numbers of premature deaths among children and mothers and increases the risk of respiratory illness and eye infections	• Gathering and preparing traditional fuels exposes women and children to health risks and reduces time spent on childcare • Modern energy is safer (fewer house fires)	• Electricity enables pumped clean water and purification • Latrines for biogas production improve sanitation

Table 1.8: continued

Target	Fundamental	Importance of energy to achieving the target	
		Contributes directly	Contributes indirectly
Access through the *primary healthcare system* to reproductive health services for all individuals of appropriate ages as soon as possible and no later than 2015	• Energy is needed to manufacture and distribute medicines and contraceptives, and to produce educational literature	• Electricity in health centres enables provision of services at night, helps retain qualified staff and allows use of equipment e.g. to sterilize instruments, refrigerate medicines	• Electricity enables access to up-to-date health education media via the internet or backup advice via radio
The implementation of national strategies for *sustainable development* by 2005, so as to reverse current trends in the loss of environmental resources at both global and national levels by 2015	• Traditional fuel use can contribute to erosion, reduced soil fertility and desertification. Fuel use can become more sustainable through substitution, improved efficiency and use of energy crops • Using more efficient energy systems and displacing fossil fuels with renewables will reduce GHG emissions	• Mitigation of increased pollution as economy grows by cleaner fuels and greater energy efficiency • Increased agricultural productivity from being able to use machinery to plough, irrigate, harvest and process crops	• National sustainability aided by greater use of indigenous renewable energy sources instead of imported fossil fuels as economy grows • Rural energy services enable non-farm-based enterprise development and processing of non-timber forest products

Source: Adapted from a draft matrix prepared by Clive Caffall, AEA Technology, 2001.

been recently estimated to cause as many as 4 million premature deaths each year, 40 million new cases of chronic bronchitis and widespread cases of other respiratory illness. Nearly 60% of the deaths are those of children under the age of five, the result of exposure to dirty cooking fuels because of lack of access to modern energy.[22] Efficient cooking stoves with a chimney to remove smoke, or biogas or solar cookers, offer clean alternatives to cooking which directly improve the health of women and children. Recent estimates of the economic loss due to the billions of cases of respiratory illness from indoor and outdoor exposure, which cause reduced productivity and low life expectancy, amount to at least US$320 billion per year, or 6% of GNP of developing countries. In China, the cost of urban air pollution alone has been estimated as 5% of GDP in 1995.[23]

Modern energy allows the provision of better healthcare services in health centres through lighting, sterilization, vaccine refrigeration and access to communications for backup medical advice and up-to-date medical literature. These facilities help attract and retain well-qualified medical staff. Modern energy also provides energy services for pumping clean water, which can help reduce the risk of waterborne diseases. Latrines for biogas production can improve sanitation where toilets do not already exist, contributing to better health.

National strategies for sustainable development   Sustainable clean forms of energy are fundamental to achieving sustainable development. Biomass, if used sustainably, can protect soil fertility and guard against deforestation and desertification. Renewable energy can contribute towards reducing greenhouse gas emissions and help reverse the current trend in loss of environmental resources at both local and global levels. National sustainability is aided by greater use of indigenous renewable energy sources instead of imported fossil fuels as the economy grows. Rural energy services make possible non-farm-based rural enterprise development and create jobs in rural areas, helping to slow down rural-to-urban migration.

---

[22] World Bank (2000a), p. 25.
[23] Ibid., p. 27.

1.3 Reducing environmental impacts

There are environmental impacts associated with the production, distribution and use of fossil fuels, and the manufacture of the associated technology. Combustion of fossil fuels gives rise to emissions of carbon dioxide (which contributes to global warming), oxides of nitrogen and sulphur (which cause acid rain and contribute to global warming), and ozone and particulate emissions (which cause smog and poor air quality, particularly from combustion of gasoline and diesel in vehicles in urban areas). Energy-efficient technologies can help reduce the amount of fuel used and emissions produced.

1.3.1 Local impacts

Indoor air pollution   As noted above, there are environmental and health impacts associated with the use of traditional biomass. Indoor air pollution can be reduced dramatically with the introduction of efficient cooking stoves with a chimney to remove smoke from the house, or by using alternative renewable energy sources such as biogas or solar cookers.

Deforestation   Again as noted above, unsustainable use of traditional biomass can be associated with deforestation, reduced soil fertility, erosion and in severe cases desertification. Careful management and sustainable use of biomass resources such as wood fuel and crops, with replacement planting, prevents environmental degradation.

Urban air quality   Emissions from transport cause poor air quality in urban areas. In industrialized countries improved vehicle efficiency, emissions standards and fuel quality standards have in general reduced emissions per kilometre, but in developing countries old and inefficient vehicles produce high levels of emissions which can lead to smog. Liquid biofuels such as biodiesel, bioethanol and biomethanol are alternatives that can be used. The ethanol programme in Brazil has been successful in encouraging production of ethanol from bagasse.

Acid rain   Oxides of nitrogen and sulphur from burning fossil fuels cause acid rain, destroying forests and crops. Use of renewable energy sources, particularly wind, small hydro and solar, instead of fossil fuels reduces damage from acid rain.

Other local impacts   The extraction, processing and transport of fossil fuels cause pollution and damage to the environment. There is also the risk associated with spillage of liquid fossil fuels and the consequent contamination of land and damage to aquatic life. Renewable energy has the advantage of being used at point of source (except for biomass in some cases) and there is little or no environmental damage caused in harnessing the energy, if managed sensibly. Renewable energy can help displace the use of dry cell batteries that are often disposed of by discarding them on the ground in rural areas, contaminating the site.

There are emissions associated with the manufacture of renewable energy technology, but thereafter, if the resources are managed and used in an efficient and sustainable way, net emissions are small or absent. For example, if biomass resources are managed carefully and used sustainably in combination with energy-efficient technologies and replanting of resources, they can be considered carbon-neutral, as the growth of biomass absorbs the same amount of carbon that is released during combustion.

The cost of energy sources should be evaluated on a life-cycle basis to take into account the long-term economic costs. Environmental and social impacts need to be included for a more realistic assessment of the life-cycle costs. Externalities need to be costed and where possible internalized, and credit should be given for additional benefits that energy sources can bring, both directly and indirectly.

1.3.2   Global impacts

The UNFCCC and the Kyoto Protocol   The 'greenhouse effect' has been a concern and a point of debate among scientists since as far back as 1827. In the late 1950s a network of monitoring stations was set up and subsequent observations showed a steady rise in carbon dioxide emissions. By 1970 the secretary-general of the United Nations was concerned enough to mention the possibility of catastrophic warming in his report on the environment. During the 1970s international attention intensified, and at the first World Climate Conference held in 1979 the World Climate Research Program was established. In 1988 the Intergovernmental Panel on Climate Change (IPCC) was formed under the auspices of the United Nations Environment Programme (UNEP) and the World Meteorological Office (WMO). The purpose of the IPCC is to produce authoritative assessments to governments on the state of knowledge concerning climate change. These reports provide the scientific underpinning for the diplomatic processes of the United Nations Framework Convention on Climate Change (UNFCCC), which was agreed in 1992 at the Earth Summit in Rio de Janeiro and came into force in 1994. The IPCC itself is precluded from making policy recommendations. Its purpose is to establish the foundation of internationally accepted knowledge upon which other forums can base their negotiations and conclusions. As a basis for action to try to achieve the aims of the Convention, the Berlin Mandate emerged from the first Conference of the Parties (COP-1), in 1995, with the first moves towards the Kyoto Protocol. A further two years of debate and negotiations ensued before the Kyoto Protocol was signed in 1997, committing industrialized countries to achieving emissions reductions to levels averaging 5.2% below those of 1990.

The ultimate objective of the UNFCCC is to achieve 'stabilisation of greenhouse gas concentrations in the atmosphere at such a level that would prevent dangerous interference with the climate system. Such a level should be achieved within a timeframe sufficient to allow ecosystems to adapt naturally to climate change, to ensure that food production is not threatened and to enable economic development to proceed in a sustainable manner.'[24] The achievement of this objective will require considerable technological innovation in combination with rapid and widespread technology transfer and implementation. This should also include transfer of information and skills development. A special climate change fund has been proposed that would aim to finance activities, programmes and measures related to climate change, in

---

[24] Article 2 of the UNFCCC.

Table 1.9: Greenhouse gases in the Kyoto Protocol

Gas	Qualifying sources	Emissions trends since the late 1980s	Lifetime (years)	GWP-100[a]	% GHG 1990, Annex I
Carbon dioxide (CO_2)	Fossil fuel burning, cement production	EU static; Other OECD increasing; EITs sharp decline	Variable, with dominant component c. 100 years	1	81.2
Methane (CH_4)	Rice, cattle, biomass burning and decay, fossil fuel production	Decline in most countries (big increase only in Canada, USA, Norway)	12.2 ± 3	21	13.7
Nitrous oxide (N_2O)	Fertilizers, fossil fuel burning, land conversion to agriculture	Varies: small increase in many countries, decline expected before 2000 in industrialized countries (but still slow increase), decline in EITs	120	310	4.0
Hydrofluorocarbons (HFCs)	Industry, refrigerants	Fast-rising emissions due to substitution for CFCs	1.5–264; HFC 134a (most common) 14.6	140/1,700; HFC 134a (most common) 1,300	0.56
Perfluorocarbons (PFCs)	Industry, aluminium, electronic and electrical industries, firefighting, solvents	Static	2,600–50,000	Average about 6,700; CF_4 6,500; C_2F_6 9,200	0.29
Sulphur hexafluoride (SF_6)	Electronic and electrical industries; insulation	Increase in most countries, further rise expected	3,200	23,900	0.30

[a] Global warming potential over a 100-year time horizon compared to that of carbon dioxide, estimated by the IPCC in its *Second Assessment Report*, Working Group I. The numbers are likely to be adjusted in subsequent assessment reports.
Source: Grubb et al. (1999).

the fields of technology transfer, capacity building, economic diversification, energy, transport, industry, agriculture, forestry and waste. The activities, programmes and measures would be required to be additional and complementary to those funded under the climate change focal area of the Global Environment Facility (GEF) and by multilateral and bilateral funding. The fund would be amassed from contributions by industrialized countries and economies in transition in the form of financial contributions and/or units of assigned amount of emissions ('carbon credits'). The new trust fund would sit within the GEF and be managed by the GEF Council under guidance from the COP (the sovereign body of the Climate Convention).

Renewable energy is just one subset of the environmentally sound technologies (ESTs) that will be needed to achieve the desired reduction in greenhouse gas emissions. One thing that ESTs need to have in common is that they should support sustainable development.

Six main greenhouse gases (GHGs) were identified during the Kyoto Protocol negotiations. These are listed in Table 1.9. It can be seen that these gases come from various sources, that their average lifetimes in the atmosphere vary, and that they have different global warming potentials (GWP) – that is, different relative impacts on global warming per molecule. Of the six GHGs, carbon dioxide accounted for over 80% of the emissions from industrialized countries of the OECD and EITs (Annex I countries) in 1990, one of the main sources being combustion of fossil fuels for power generation. For this reason carbon dioxide will be the focus of discussions in this book.

If we continue to rely so heavily on fossil fuels, in particular oil (as predicted in Figure 1.2), the implications for increased carbon dioxide emissions are clear. Table 1.10 shows that world carbon dioxide emissions levels are predicted to rise 31% from 1997 levels by 2010 and 60% by 2020. Developing countries are expected to account for the majority of this increase, with emissions almost level with those of OECD countries in 2010 and passing them by 2020. This increase in emissions levels is worrying, given that the evidence for a link between carbon dioxide emissions and global warming is now stronger than ever. The recent report from Working Group I of the IPCC con-

Table 1.10: Global carbon dioxide emissions by region, 1990–2020 (Mt CO_2)

Year	World[a]	OECD	EIT	Developing countries
1990	20,878	10,640	4,066	6,171
1997	22,561	11,467	2,566	8,528
2010	29,575	13,289	3,091	13,195
2020	36,102	14,298	3,814	17,990

[a] Excluding international marine bunkers.
Source: WEO (2000).

cluded that 'the balance of evidence suggests a discernible human impact on global warming.'[25]

Agreements on GHG emissions   The UNFCCC recognized the difference between the situations of industrialized and developing countries in respect of their contribution to carbon dioxide emissions and their capacity to address GHG emissions levels in the short to medium term. The result was that countries were split into two groups:

- *Annex I* – industrialized countries of the OECD and EIT (e.g. countries of the former Soviet Union and central and eastern Europe): these countries have agreed to adopt emissions limitations or reductions commitments and take the lead in climate protection;
- *non-Annex I* – other countries not yet able or willing to commit to emissions limitations or reductions (mainly developing countries).

The signing of the UNFCCC in 1992 constituted a recognition by signatory countries that climate change caused by emissions of GHGs needed to be investigated further and addressed. The UNFCCC included two main agreements on the reduction of GHG emissions, one non-binding and one binding:

- Annex I parties proposed a non-binding agreement to aim to reduce GHG emissions to 1990 levels by 2000;

---

[25] IPCC (2001).

- all parties to the Convention were bound by Article 4.1 to adopt several actions either to reduce or to prepare to reduce emissions.

After the Earth Summit in Rio in 1992, subsequent meetings of the parties sought to adopt a stricter regime on emissions limitations and reductions. At COP-3 in Japan in 1997, the Kyoto Protocol was signed and quantified emissions limitation and reduction commitments (QELRCs) were agreed upon.

Kyoto mechanisms In order for the Kyoto Protocol and its associated commitments to be agreed and signed by Annex I countries, mechanisms ('Kyoto mechanisms') were included in the Protocol which enable emissions credits to be traded between countries. This allows countries to meet some of their emissions limits or reductions by actions outside their domestic situation. The argument for the Kyoto mechanisms was that GHG reductions would thus be met in a least-cost way. The process leading up to the Kyoto Protocol was very complicated, with different countries and groups of countries arguing from different points of view. But, in general, developing countries have argued that the Kyoto mechanisms are a way for industrialized countries to evade their commitments to limiting GHG emissions. To address this objection, it was stipulated that Kyoto mechanisms must be 'supplemental' to domestic action.[26] Some countries have called for a cap to be placed on the amount of credits that can be realized from action taken in other countries to try to ensure that a substantial level of action is taken domestically.

Three mechanisms are mentioned in the Kyoto Protocol: international emissions trading; project-based joint implementation between Annex I countries; and the Clean Development Mechanism between Annex I and non-Annex I countries.

1) International emissions trading: The idea for international emissions trading of GHGs was based on several models, including the successful experience of the US system of tradable permits for sulphur dioxide

---

[26] UNFCCC (2001).

emissions. The idea was that each country would be given a limit of GHG equivalent emissions, and then emission permits could be traded domestically between companies or internationally between governments. Surplus emissions credits have been created in the EITs, and most notably in Russia, as economic recession has caused GHG emissions to fall dramatically, leaving them with surplus credits to trade. The enormous quantity of these surplus credits – or 'hot air', as they were termed – is a matter of concern, as they open up the possibility of some Annex I countries positioning themselves to buy up the majority of them, thus relieving them of the need to take much or any domestic action on emissions reductions. This led to a call for a 'cap' to be placed on the amount of traded credits that could be used to achieve a reduction target. Article 17 of the Protocol contains a phrase that trading should be 'supplemental' to domestic action to meet reduction requirements, although at the time of writing the level to be met by domestic action is not indicated.

2) Joint implementation: Article 6 of the Kyoto Protocol identifies the mechanism known as joint implementation (JI), which allows Annex I countries to meet part of their targets through projects to reduce emissions in another Annex I country.
   JI projects are likely to include:

* *sources* – projects using technology which reduces sources of GHG emission (e.g. energy-efficient technology; fuel switching to less carbon-intensive fuels such as natural gas or renewables); and
* *sinks* – projects which enhance carbon sinks (e.g. forestry projects and carbon sequestration).

During the first commitment period, 2008–12 (and subsequent commitment periods), credits called emissions reduction units (ERUs) are collected by the project investor(s) and can be used by them to offset their own reductions targets or potentially traded under the emissions trading mechanism.
   There has been some debate over the merits of JI, and in particular the complexity and methodology of monitoring and evaluating the

emissions reductions resulting from JI projects. Article 6 states that JI activities should be 'supplemental' to domestic actions, but current usage of the term is not defined clearly. Emissions reductions are also required to be 'additional' to that which would have otherwise occurred. Again there is difficulty in agreeing a precise definition of this term. Setting baselines against which to measure the reduction in emissions is a complicated issue. A pilot phase of JI without any crediting of emissions reductions was established at COP-1 in Berlin in 1995. It was called 'activities implemented jointly' (AIJ). The aim of AIJ was to gain experience in how to monitor and evaluate projects and set realistic baselines against which to measure sinks or reductions in emissions of GHGs.

3) The Clean Development Mechanism (CDM): The stated purpose of the CDM is to help developing countries achieve sustainable development, thus contributing to the ultimate objective of the Convention, and to assist Annex I parties in achieving compliance with their specific emissions reduction commitments. At the time of writing the CDM is still under negotiation; it will be similar to JI, but between Annex I countries and developing countries rather than between Annex I countries. Because the transfer of credits will be from developing countries which do not have agreed emissions reduction limits, the issues of accurate monitoring, evaluation and baselines are of even greater concern than for JI.

There are four main features that distinguish the CDM from JI:

- multilateral control is stronger (the CDM will be supervised by an executive board);
- funds will be provided to cover administration costs and the costs of particularly vulnerable developing countries to adapt to climate change;
- projects must show sustainability benefits to host countries;
- projects will be allowed early crediting from 2000 onwards (see below).

CDM project activities should in theory benefit developing countries via investment and technology transfer, and generate 'certified emissions reductions' (CERs) which Annex I parties may use to contribute to compliance with their QELRCs. CDM project emissions reductions will be certified on the basis of certain criteria being met:

- CDM projects must show real, measurable, long-term benefits relating to climate change;
- they must ensure sustainable development benefits in the beneficiary country;
- there must be voluntary participation in the project; and
- emissions reductions must be additional to any that would occur in the absence of the project (this has been termed 'additionality').

As with emissions trading and JI, the CDM projects have to be 'supplemental' to domestic action.

It is intended that the CDM will assist in the funding of certified project activities, but not itself be a fund. Both private and/or public organizations can take part in CDM projects. To ensure that the CDM is transparent, efficient and accountable, there will be independent auditing and verification of project activities.

It is stated in Article 12.10 of the Kyoto Protocol that CERs from the year 2000 up to the beginning of the first commitment period (2008) can be used to achieve compliance in the first commitment period (2008–12); this is known as 'early crediting'. It has been agreed that Annex I parties should refrain from using nuclear projects under the CDM and that in the first commitment period sinks should be restricted to afforestation and reforestation projects only.[27] Large hydro and clean coal projects could still be allowed. The structure and operation of the CDM, along with the need for clear definitions of the criteria to be used for selecting projects eligible under the mechanism and the methodologies for monitoring, evaluation and crediting, have sparked much debate. What is eventually agreed on these points will affect investment in developing countries and the technology that is transferred (these issues are covered in more detail in Chapter 2). Lessons learnt from AIJ on setting baselines, monitoring and evaluation will be a

---

[27] Ibid.

useful guide for discussion and agreement on how the CDM is to be operated.

To help speed up the implementation of CDM projects and take advantage of early crediting, some simplified procedures are to be agreed for small-scale projects, which include small-scale renewable energy projects of up to 15 MW capacity or equivalent. This will benefit small-scale projects by reducing transaction costs. To achieve this prompt start, it was agreed that the CDM executive board was to be chosen at COP-7 in 2001 and would recommend simplified rules for small-scale projects at COP-8 in 2002.[28]

It is recognized that current approaches to energy use are unsustainable, and there is a significant potential for both renewable energy and energy efficiency to be integral elements in any future plans to respond positively to the issues of sustainable development. Sustainable development is a continuing theme throughout the UNFCCC and the Kyoto Protocol. The CDM has a dual role in that it aims to transfer sustainable clean energy technology to developing countries to contribute towards developmental goals at the same time as producing emissions credits for the industrialized countries to help them comply with their emissions reduction targets. Renewable energy is well suited to take on that dual role. For the transfer of renewable energy to be successful and sustainable, appropriate policies, legislation and regulation for renewables will need to be put in place in the host countries, and local capacity will need to be built up to select, install, maintain and adapt systems to local conditions. Progress on the Kyoto Protocol made at COP-6bis in Bonn in July 2001 and at COP-7 in Marrakech later the same year has raised hopes that it will be adopted; however, even if the Protocol does not go ahead in its present form, it is none the less likely that technology transfer along the broad lines set out in the Protocol is likely to form part of any future climate change agreements.

---

[28] Ibid.

Chapter 2

Transferring Technology to Developing Countries: Key Actors and Roles

2.1 Developments in technology transfer

The classic image of technology transfer was large-scale public investment based on foreign technology and soft loans,[1] with minimal knowledge transfer and domestic capacity building. In the 1950s and 1960s this approach was the norm: technology was parachuted in with little attention given to building domestic capabilities to operate and maintain the equipment. The oil crisis of the 1970s changed the situation dramatically as countries dependent on oil imports searched for domestic energy substitutes and underwent a shift from supply-side management to demand-side management in an attempt to gain greater energy security. Where possible, national oil companies were developed and contracts between governments and international oil companies became much more stringent in respect of the demands made on the oil companies, with a greater emphasis put on training local staff and purchasing locally manufactured equipment. Following this, new ventures in small-scale renewable energy technology and energy-efficient technology began to emerge, which began to influence views on technology transfer. These smaller-scale, dispersed technologies needed a greater number of locally trained people to provide installation and maintenance support, thus raising awareness of the need to transfer skills as well as technology. As environmental and social concerns became more important, nuclear and large-scale hydro projects were called into question, and greater attention began to be paid to other renewable energy sources, such as smaller-scale hydro (including run-of-the-river hydro), geothermal, solar, wind and biomass. In addition, less carbon-intensive fossil fuels such as natural gas began to be exploited.

---

[1] Loans issued on more favourable terms than commercial lending rates (e.g. with lower interest rates and over the long term).

Multilateral agencies started to include technical assistance with training, research and development activities as standard components of project financing packages. The emphasis shifted away from isolated hardware package deals to more integrated, process-oriented approaches, including proper incentive structures for relevant actors.[2] The oil shocks of the 1970s and early 1980s, coupled with rapid population growth in developing countries, increased the demand for indigenous energy sources and technologies in developing countries and thus the need for renewable energy technology transfer.

In the 1990s market globalization accelerated and the availability of private capital on a global scale increased, as did competition between global vendors of technology. Market restructuring through liberalization and privatization started to spread to some developing countries. With privatization, the role of governments in the technology transfer process changes. Prior to privatization and restructuring, governments (through their national utilities) played an active role as recipients in the process. Since privatization and restructuring they have begun to concentrate more on regulating trade and promoting enabling policies (economic incentives; legal aspects of innovation policies). Governments now have a key role in facilitating the diffusion of technology through creation of an adequate institutional infrastructure with high-quality engineering education, promotion of R&D activities, adequate industrial standards and flexible market mechanisms.[3]

2.1.1 Defining technology transfer

Technology transfer can be defined as the diffusion and adoption of new technical equipment, practices and know-how between actors (e.g. private sector, government sector, finance institutions, NGOs, research bodies, etc.) within a region or from one region to another. Most transfer is North–South, but more South–South transfer needs to be encouraged. The definition of technology transfer seems straightforward, but the process is complicated. Successful technology

---

[2] IPCC (2000), ch. 10.
[3] Ibid.

transfer needs attention to the following aspects: affordability, accessibility, sustainability, relevance and acceptability. Particular attention needs to be paid to commercial management; market development (meeting the energy requirements of the people); economic competitiveness; and technical adaptation to local conditions.

'Technology' should be regarded not only as the equipment, but also the information, skills and know-how which are needed to fund, manufacture, install, operate and maintain the equipment. 'Transfer' should be regarded as putting the technical concepts into practice locally in a sustainable framework so that local people can understand the technology, use it in a sustainable manner and replicate projects to speed up successful implementation. Transfer of technology also includes improvements to existing technology (e.g. improved cooking stoves) and adaptation of technology to local conditions and requirements. Technology transfer should assist local people in developing the skills to choose appropriate technology, adapt it to local conditions and energy service requirements, and integrate it with existing indigenous technology.

Technology transfer can be described as 'vertical' or 'horizontal'. Vertical transfer refers to the point-to-point relocation of new technologies via investment, often to a target group. There is no transfer of knowledge and skills to local manufacturers. The activities of the company are extended without risk to technology ownership. For example, a large multinational might set up a factory to manufacture its technology in a developing country, as this brings down the costs of operation. The factory would be wholly owned and operated by the company. The management and technical staff might be expatriates, but the general workforce would be cheap local labour. The company would not expose its technical designs to potential competitors and would control the quality of production. Horizontal transfer describes the long-term process of embedding technology within local populations and economies, including technical and business training and financial management. This may lead to the local manufacture and ownership of the technology, via the establishment of a joint venture (JV) between a foreign and a local company.[4] This approach can make it more difficult

---

[4] Forsyth (1999), ch. 3.

for foreign companies to protect their designs and control the quality of products, as the local partner will be manufacturing the technology; however, it can lead to a more sustainable situation as skills and knowledge are built up in the developing country that can help in the installation, operation and maintenance of the new technology.

In general, technology is owned by companies rather than governments; therefore, successful technology transfer will not happen without the involvement of companies. Companies play a key role before, during and after the transfer process. In fact, if a joint venture is set up between a foreign company and a local company, then technology transfer is a continuous process as further technical developments take place and skills and know-how are periodically updated.

However, commercial enterprises are driven by the need to make profit. JVs can prove to be costly, and many companies fear the loss of intellectual property rights. But many developing countries are now beginning to insist on the formation of JVs by international investors as a way of encouraging horizontal technology transfer. In the long term, horizontal transfer should be a more sustainable approach to technology transfer from a developing-country perspective; however, this is not necessarily so from the perspective of a company, whose primary driver is profit. Yet without the engagement of private companies, little technology transfer will occur.

Historically, technology and knowledge have been fiercely guarded by both governments and companies, given that an advantage in either can lead to increased economic or military power. The encouragement of technology transfer by countries is consequently a relatively new practice, seen only since the second half of the twentieth century.[5] Contributing factors to the change in patterns of technology transfer and cooperation include the development of multilateral organizations (e.g. the United Nations) and transnational corporations, and the development of communications and intellectual property rights legislation.[6]

Agenda 21, Chapter 34, points out that 'Technology co-operation involves joint efforts by enterprises and governments, both suppliers of

---

[5] Siddiqi (1990).
[6] IPCC (2000).

Figure 2.1: Technology transfer, linear model

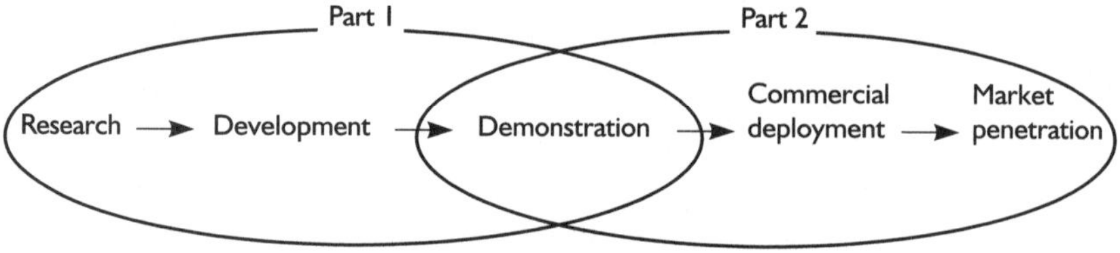

Part 1 ———— Part 2

Research ⟶ Development ⟶ Demonstration ⟶ Commercial deployment ⟶ Market penetration

Source: Adapted from Dixon (1999).

technology and its recipients. Therefore, such co-operation entails an interactive process involving government, the private sector, and research and development facilities to ensure the best possible results from transfer of technology.'[7]

Various stages can be identified in the process of technology transfer. A range of different models is used to describe these stages. For example, there is a linear model which identifies five stages from research to development, to demonstration, to commercial deployment and finally to market penetration. These stages can be grouped into two overlapping parts, as shown in Figure 2.1, with demonstration fitting into both. During the first part, R&D activities are carried out with market needs in mind, the results of R&D are effectively exploited and disseminated, and good communication networks are set up between industry and research. In the second part, activities reduce non-technical barriers to the technology, carry out demonstration appropriate for target markets, disseminate demonstration results, provide support to replicate projects, raise market awareness and knowledge of up-to-date information, and create market enablement and access. During the whole process there needs to be responsive strategic development, effective dissemination tools, proactive and dynamic interactions, and monitoring and evaluation of activities.[8]

Another model is the five-stage feedback model which goes from assessment to agreement, to implementation, to evaluation and adjustment, to replication, and them back round to assessment (see Figure 2.2).[9]

---

[7] UN (1993).
[8] Dixon (1999).
[9] IPCC (2000).

Figure 2.2: Technology transfer, feedback model

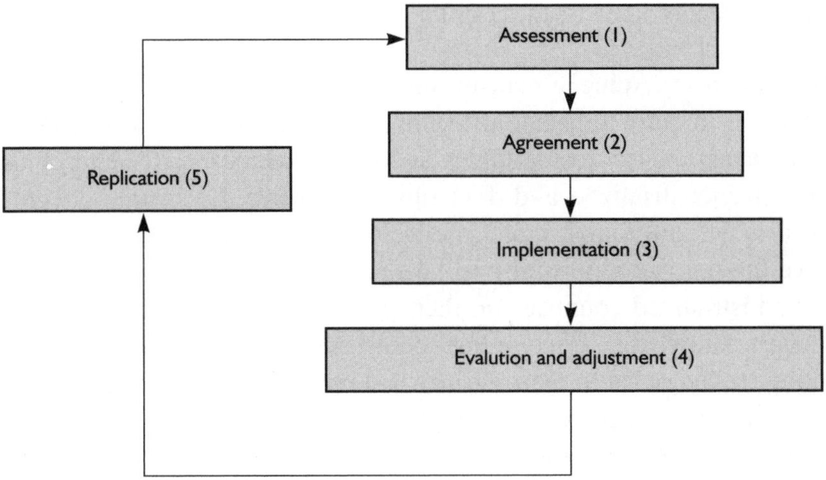

Source: Adapted from IPCC (2000), p. 57.

Whichever model is used, different actors will get involved with the technology development and transfer process at different stages. An essential part of both models, vital for stepping up the speed and capacity of renewable energy technology transferred to developing countries, is replication of best practice and successful projects.

2.1.2   Technology transfer in the context of sustainable development and climate change

Technological development and innovation for environmentally sound technology (EST) are needed globally. Transfer of EST to developing countries and economies in transition is crucial if sustainable development is to be achieved and climate change is to be addressed. Environmentally sound technologies include technologies 'which protect the environment, are less polluting, use all resources in a more sustainable manner, recycle more of their wastes and products, handle residual wastes in a more acceptable manner than the technologies for which they were substitutes, and are compatible with nationally determined socio-economic, cultural and environmental priorities'.[10] Renewable energy technologies fit these criteria well.

Development Achievement of sustainable development on a global scale will require substantial technological changes to more efficient and cleaner energy technologies and improved patterns of energy use in both industrialized and developing countries. Economic development is currently most rapid in developing countries. These countries have the unique opportunity to take advantage of the lessons learned by industrialized countries on their path towards development. Using modern knowledge, developing countries can choose to leapfrog polluting technologies to their cleaner and more efficient successors, thus avoiding past unsustainable practices and adopting environmentally sound technologies, practices and institutions. Renewable energy has an advantage over fossil fuels and grid electricity in that it can serve remote populations in rural areas that are difficult to access, as the energy resources can be harnessed locally, removing the need for transportation or distribution of fuels.

For developing countries to take full advantage of renewable energy technologies and other ESTs they need assistance in building local knowledge, technical and business management skills and appropriate institutions and networks. It is also important to ensure that renewable energy technology is adapted to the local environmental conditions and requirements of the people and businesses being served; and that local people are able to make an informed choice about which technologies best meet their requirements, have the least impact on the environment and are best affordable. As different societies have different priorities and requirements for sustainable development, it is important that renewable energy systems be selected and tailored to those requirements and priorities and fit in with each community's social and cultural context. The expectations of the community need also to be managed carefully, so as not to promise what cannot be afforded or delivered in practice.

---

[10] UN (1993).

In the early to mid-1990s development agencies that had helped fund large power projects did not see the development benefits hoped for. This made them look more closely at socio-economic issues related to gender, poverty alleviation, health and education. In the process, the focus on energy has been diminished to the extent that the energy element in providing such sorely needed services has often been neglected. There are now signs that some multilaterals and bilaterals are beginning to realize that this balance needs to be redressed and are looking at how to integrate energy back into their activities. For example, the World Bank has recently launched an energy renewal strategy, which includes a business strategy for energy focusing on four areas: direct poverty alleviation; environmental sustainability; macroeconomic and fiscal stabilization; and governance and private sector development. The strategy will look closely at the linkages between energy and poverty, and also between energy and environment.

Ensuring that sustainable clean energy is considered when services such as education, clean water, etc. are being planned in developing countries might best be achieved by integrating energy as a theme in other sectors, rather than re-creating it as a sector in its own right. This might help to avoid setting energy targets which can detract from the real requirements. For example, meeting an energy target of '100% villages electrified' in a particular region may in itself achieve very little, especially if the target is seen to be reached and assistance is withdrawn without looking closely at the benefits (or lack of them) it has brought to the communities. Further steps need to be taken to address affordability of electricity supply, access to credit, how use of the electricity can be planned to provide community services, how the electricity could contribute towards enterprise development, etc. As described in Chapter 1 above, energy is essential for economic growth and can contribute directly and indirectly to poverty alleviation, helping to achieve international development targets; but it is not energy in itself that is desired or needed so much as the things that it renders accessible, such as improvements in people's health through reducing indoor air pollution and provision of better healthcare facilities; the creation of local jobs and income-generating activities; and improved prospects for education via access to educational media, information and communications. So sustainable energy needs to be considered as one of the issues integral to planning services in each sector.

Climate change In the context of climate change negotiations, the transfer of EST to developing countries and EIT is recognized as being essential to help them industrialize while limiting their emissions. Technology transfer is an objective of both the UNFCCC and the Kyoto Protocol.

Throughout the UNFCCC there are references to technology transfer. Article 4.1(c) of the UNFCCC commits all parties to the Convention to 'promote and co-operate in the development, application and diffusion, including transfer, of technologies, practices and processes that control, reduce or prevent anthropogenic emissions of greenhouse gases'. Article 4.5 invites industrialized country parties in particular to 'take all practical steps to promote, facilitate and finance, as appropriate, the transfer of, or access to, environmentally sound technologies and know-how to other Parties, particularly developing country Parties, to enable them to implement the provisions of the convention'. Article 4.9 specifically mentions the needs of least developed countries that are least likely to be able to afford new technology: 'the Parties shall take full account of the specific needs and special situations of the least developed countries in their actions with regard to funding and transfer of technology.' Article 11.1 defines a mechanism for the provision of financial resources on a grant or concessional basis, including for the transfer of technology.

Agenda 21 states in Chapter 34 that access to and transfer of EST should be promoted 'on favourable terms, including on occasional and preferential terms, as mutually agreed, taking into account the need to protect intellectual property rights as well as the special needs of developing countries for the implementation of Agenda 21'.

Technology transfer is mentioned in several places in the Kyoto Protocol. Article 3.14, on commitments, acknowledges the need to minimize the adverse impacts of climate change in developing countries and notes that among the 'issues to be considered shall be the establishment of funding, insurance and technology transfer'. Article 10(c), on sustainable development, talks about the promotion, access and transfer of EST to developing countries. All parties shall co-operate in the promotion of effective modalities for the development, application and diffusion of, and take all practicable steps to promote, facilitate and finance, as appropriate, the transfer of, or access to, environmentally sound technologies, know-how, practices and processes pertinent to climate change, in particular to developing countries, including the formulation of policies and programs for the effective transfer of environmentally sound technologies that are publicly owned or in the public domain and the creation of an enabling environment for the private sector, to promote and enhance the transfer of, and access to, environmentally sound technologies.

JI and the CDM are the two mechanisms under the Kyoto Protocol by which technology transfer could be encouraged to EITs and developing countries respectively. The Protocol does not make any explicit link between the Kyoto mechanisms and technology transfer. However, it does acknowledge that technology transfer is increasingly being integrated into policy debates about investment, and that the problems of managing investment for better development are similar to those for increasing technology transfer.

Meeting climate change objectives alone is not a strong enough driver to ensure the transfer of renewable energy technology from industrialized to developing countries, as both groups are motivated more by objectives of economic development and international competitiveness. If industrialized country governments are legally bound or otherwise forced to reduce emissions, have targets and commitments to reach, and have mechanisms in place (such as the CDM) that encourage the transfer of EST, the environment is likely to become a more important driver for energy technology transfer. The CDM was created as a multilateral mechanism to assist developing countries in achieving sustainable development while at the same time allowing industrialized countries to comply with their GHG reduction commitments at least cost.

The CDM has great potential to create partnerships between the private sector in industrialized countries and between governments and the private sector in developing countries, which could facilitate the transfer of renewable energy technology. It is well recognized that one of the largest constraints on renewable energy technology transfer is ineffective channelling of funds for investment in renewable energy technology (RET), rather than a lack of private capital itself. CDM could help harness private capital and channel it to RET in developing countries. Chapter 3 looks at investment in RET and the potential role of the CDM.

At the time of writing, the details of the CDM are still under negotiation, and as a consequence details of current agreements will most certainly be out of date by the time this book is published; therefore, the reader is referred to the UNFCCC website for the latest agreements on the CDM and other Kyoto mechanisms.[11] The final form taken by the CDM will be important in determining the impact it has on speeding up renewable energy technology transfer to developing countries in the future. The following factors are among those that have been considered in the process of developing the CDM, and have sparked much debate and disagreement.[12]

❑ Environmental additionality: Projects need to create emissions reductions beyond those which would have occurred in a business-as-usual situation. There is a paradox in trying to meet emissions reductions through least-cost projects, in that the more cost-effective a CDM project is (i.e. the lower the investment needed in relation to the credits received), the more uncertain its additionality will be. This is because only a small shift in market conditions could make the project economic enough to go ahead without credits, so that it could have happened without CDM investment; in these circumstances it would not be additional. However, there may be other barriers (apart from the economics of the project) which would have prevented it from being implemented. In reality there may be no way of knowing if a project would have gone ahead without a CER. If the CDM is not implemented strictly and enforced easily, there is a danger that it could weaken action in the industrialized countries without any offsetting real reductions in the emissions of host developing countries. Given these difficulties, the CDM could be viewed as a mechanism that merely accelerates clean

---

[11] The website address is <http://www.unfccc.de>.
[12] Based on information in Grubb et al. (1999).

investments that might well have occurred eventually, implying that emissions reduction should be considered additional only for a limited duration. A recent study for the WWF highlights the potential problem of 'free-riders' carrying out projects under the CDM that would have occurred anyway.[13] The presence of a small proportion of free-riders is inevitable and acceptable as long as the CDM helps to achieve the ultimate objectives of the Climate Change Convention, for example, speeding up the development and adoption of technologies that could underpin a global transition away from carbon-intensive fuels and contributing to sustainable development. However, unless the eligibility criteria for projects, baseline setting and additionality testing are well thought out and strictly implemented, the potential for free-riders could threaten the environmental integrity of the Kyoto Protocol, leading to increased carbon emissions. The WWF study estimates that if CDM projects were limited to non-hydro renewables projects only, the potential for free-riders would be dramatically reduced. It concludes that it is essential that policy-makers devise and adopt a CDM regime that effectively encourages legitimate projects, while at the same time rigorously screening out non-additional activities.

❑ Project eligibility: There has been, and continues to be, much debate as to what types of project will be eligible under the CDM. The main points of discussion, and the interim agreements reached, including measures to fast-track small-scale renewable energy projects, have been outlined in Chapter 1 above. All projects will need to fulfil certain sustainability criteria which are yet to be agreed in detail. These criteria (and their indicators) might include the following:[14]

- technology transfer (initiation, imported/local technology, energy/ cultural requirements met, affordability, training, quality control, participatory planning, subsidy, access to credit/stability of income);
- poverty alleviation (economic impact, freed time, access to energy, social benefits, health and safety, education, access for poor, environment);

---

[13] Bernow et al. (2000).

[14] Begg et al. (2000), annex 4.

- capacity building (enhancing human resources and raising public awareness, strengthening existing institutions, expanding national and regional resources of strategic data and information, stimulating policy reform to create enabling environments);
- environmental effects (land use, water use and pollution, waste disposal, air pollution, climate change, biodiversity, resource depletion, dispersion of toxic substances);
- country priorities (reduction of pressure on biomass resources, economic improvement and poverty reduction, social infrastructure improvements in health care and education, reduced dependence on imported fuel, reduced air and water pollution, conservation, security of energy supply, rural electrification).

❑ Crediting and certification: There is pressure from investors seeking some reassurance of the credits their investments will reap. If project developers get the baseline certified before the start of the project, they should be able to anticipate what credits they will receive if the project is successful and be able to estimate the risk that they will be taking. But actual credits can not be verified and given until the project is in place and operational. There will need to be separate bodies involved in investment, validation, certification and verification, to minimize the possibilities of fraud and corruption. The distribution of credits between project partners will also be difficult, especially if developing countries are contributing towards the project investment. Credit sharing and tradability of credits will need to be defined clearly.

❑ Permanence: There has been much debate over the permanence of carbon credits for, say, forestry projects as the lifetime of the trees is not guaranteed. For example, how should crediting be dealt with if plantations are subsequently illegally logged or burnt in forest fires, through no fault of the plantation owner – circumstances in which the carbon would be released back into the atmosphere? It has been proposed that a contingency fund be set up to help deal with such problems, but no details have been agreed as yet. At COP-7 it was agreed that a new removal unit (RMU) would be used to represent sinks credits generated by Annex I countries that cannot be banked for future commitment periods.

❑ Distribution of investment among developing countries: There is concern regarding the geographical distribution of CDM projects. African countries are particularly worried that the CDM might widen the gap in international investment flows between them and other developing countries. They have proposed that CDM projects should be allocated evenly among different groups and regions within the G77. However, quota systems are often problematic and slow to implement. It has been agreed that public funding for CDM projects from Annex I parties must not result in the diversion of overseas development assistance (ODA) from OECD to developing countries, and is to be separated from and not counted towards the financial obligations of the Annex I parties.[15]

❑ Sources of finance: It is stated in the Kyoto Protocol that private and/or public entities may participate in CDM projects. This raises the question of 'financial additionality': i.e. is the financing additional to what the country would have received anyway? Developing countries are concerned that current government ODA investments will be channelled into CDM projects and therefore would not be additional or new investments. The stipulation noted above, that ODA is not eligible for CDM projects, is in part designed to avoid this eventuality. Although investment cannot be restricted to private organizations, it is possible to restrict crediting to them, clarifying the complementary roles of ODA and the CDM. ODA can remain a channel to help developing countries adopt more sustainable practices (capacity building, infrastructure development and investments in demonstration projects of innovative technologies), while the CDM can aid the greening of foreign private investment which is crucial for technology transfer and development.

❑ Incentives to invest: In order to encourage private investment in CDM projects, there has to be pressure from domestic emissions-reduction legislation. Without this, there is no need for the private sector to act.

❑ Governance: The Kyoto Protocol states that an executive board will be in charge of governing the CDM. It can be argued that its members

---

[15] UNFCCC (2001).

should include representatives from both developing countries and industrialized countries, as CDM projects will most likely be between these countries; representation from EIT, as they have vested interests in ensuring the CDM has stringent rules; a corporate representative, as private investment is crucial for the CDM; and a representative from the environmental NGO community. However, the bigger the board, the longer it will take to make decisions. At COP-6bis it was agreed that the board should consist of one member from each of the five UN regional groups, plus one member from the Association of Small Island States (AOSIS); two members from Annex I parties; and two members from non-Annex I parties.[16] At COP-7 ten members and ten alternatives were appointed to the CDM executive board.

❑ Administration and adaptation charges: A share of proceeds from certified project activities is intended to be used to cover administrative expenses and to assist vulnerable developing countries to adapt to the adverse effects of climate change. There have been proposals that the adaptation fund will be funded from a share of proceeds of the CDM (2% of CERs generated) and contributions from Annex I parties (the latter in particular during the start-up period).[17] The proceeds would be collected after project reduction has been certified and CERs issued.

2.2   Key actors and roles

2.2.1   Key actors

Technology transfer happens as a result of actions taken by different individuals or organizations that participate in the process. These can be referred to as actors, players or stakeholders. The key actors include policy-makers, legal and regulatory bodies, development agencies, financiers, utilities, manufacturers, suppliers, developers, installers, consultants, academic institutions, NGOs, community groups, recipients and users of the technology. There are various ways in which the technology transfer process can take place, involving different actors

---

[16] Ibid.

[17] New proposals by the President of COP-6, 9 April 2001; available at <www.UNFCCC.org>.

taking on different roles. For example, the transfer process can happen through joint ventures, foreign direct investment (FDI), government assistance programmes, direct purchases, joint research and development programmes, franchising and sale of turn-key plants.[18] In general the technology transfer process is initiated by one of four main actors: government, international finance institutions (IFIs), private sector or civil society (e.g. NGOs and communities). The government sector usually drives the transfer process to fulfil certain political and social goals (e.g. economic and social development, environmental improvements). The private sector usually has commercial goals to satisfy. International finance institutions include development agencies and regional development banks, so can have both social goals and commercial goals. Civil society typically drives the process to satisfy local demands, for example, creation of jobs and income-generating activities, improvements in education, health care and housing services, and better access to communications and household appliances (e.g. CB radio, radio/cassette, TV and telephone). For a community to drive the technology transfer process effectively, it needs an effective and high degree of collective decision-making, defined goals and, often, a champion within the community.

The most effective way in which to transfer a technology will depend on the local governance structure, the political and legislative context, the actors involved and the degree of success of the technology in similar applications elsewhere. Table 2.1 summarizes some of the different actors and their roles in the process of renewable energy technology transfer to developing countries.

The processes of technology transfer for large-scale fossil-fuel-based energy systems such as oil, coal and gas are well established and rely heavily on a few large private-sector actors (e.g. multinational companies and private investors). In contrast to this, renewable energy systems (RES) are usually small-scale decentralized technologies, often with a large number of actors (ranging in size from individual recipients of the technology through small and medium-sized enterprises to large

---

[18] IPCC (2000). This report calls the different ways in which the technology process can happen 'pathways'.

Table 2.1: Actors and roles in renewable energy technology transfer

Actor	Characteristic role	Comments
Central government	Financing construction via national budget and negotiation of international aid programmes in its country	Funding can overlook prevailing market conditions; political drivers can change; budgetary priorities and conditions may change
State/provincial government	Financing construction via delegation of national budget	Competition for funds from other areas. Possible lack of expertise in finance, management, etc.
Development agencies	Financing construction via national budget	Competition for funds from other areas
Local (community) government	Smallest formal structure concerned with infrastructure and development issues	Lack of relevant expertise in finance, construction, management, etc.
Villages	Non-formal government structure, concerned with local developments and governance	Not a formal legal entity. Limited skills, ability to plan and execute
NGOs	Financing through third-party contributions	Limited finance available
Private companies	Investment in projects offering adequate return on investment	Driven by market conditions and economics, not by social development concerns
Cooperatives	Grouping of producers/consumers with common goals. Provide an institutional structure through which commodities and services can be delivered and paid for	Lack of skill and expertise. Moving towards reliance on simple bottom line profit/loss or cost/benefit criteria. Low access to capital
Families	Basic unit of society/production in agricultural societies. The potential users of renewable energy systems	Low access to capital and skills. Social structures/culture may cause problem
Lending institutions	Investment in projects offering adequate return on investment	Usually require large projects and national government involvement
Educational institutions	Investment in manpower and human capital	Access can be restricted for women or minorities or the very poor
National utilities	Usually single-purpose with a very focused mission statement	Single-mission focus can cause 'blinkered' view. May have vested interests, e.g. in distribution system

Table 2.1: continued

Actor	Characteristic role	Comments
Rural energy service companies (RESCOs)	Regional and local focus, providing access to energy services, via purchase, lease or fee for service	Need to reach critical mass to support management and technology repair/replacement
Regional businesses	Looking for sustainable levels of business with adequate added value	Need to reach critical mass to support activities such as custom design and installation
Local business	Often headed by a local entrepreneur. Provides access to appropriate finance and technology locally	Needs institutional support plus technical backup (see also families). Often low access to capital

Source: Adapted from Williams and Bloyd (1997) and ETSU (1999a).

multilateral agencies with development or environmental goals) who have an influence on the development of the system. This is not to say that large multinational companies are not interested in or getting involved with renewable energy. Oil companies such as Shell and BP have renewable energy businesses, mainly focused on PV technology at present. The motivation for these companies in diversifying into renewables is partly a result of a longer-term vision to become energy companies rather than just oil companies, providing a more secure future for their businesses as oil resources become more difficult and costly to exploit. This is coupled with growing international pressure to reduce carbon emissions, national legislation to reduce a range of emissions and pressure from civil society to be more environmentally conscious and green.

Figure 2.3 shows the range of actors that can have an impact on the development of renewable energy systems and some of the channels of influence. The figure is drawn from a developer-centred view for a grid-connected system. The developer could be a local organization, a foreign organization or a joint venture. For some renewable energy projects such as those involving wind, solar and hydro, there would be no need for fuel supply contracts and no fuel costs. For stand-alone renewable energy systems, there would be no need for a power purchase agreement with the local distribution company, although it may be necessary to agree a franchise with them for supply of energy services

Figure 2.3: Potential project participants (developer-centred view)

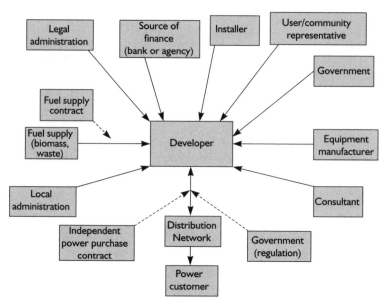

Source: Adapted from ETSU (1999a).

to households in their region. Not all actors will be involved in the transfer process; the particular combination of actors in each case will depend on the way in which the transfer process is arranged.

Renewable energy projects are site-specific, being influenced by the local resources available and the energy services required. These factors will in turn impact on the type of scheme most appropriate for the area and the actors involved. Biomass projects can involve more actors and be more complex because of the need to secure a reliable fuel supply at a reasonable cost (if not free). It is extremely important to assess both energy requirements and the renewable energy resources available (e.g. biomass supplies, wind regime, solar radiation or hydro resource) before planning any renewable energy scheme.

Given the number of actors involved in the implementation of renewable energy systems, coordination of the technology transfer process is particularly important to assist the successful flow of technology. It is also important to encourage partnerships among actors, particularly to secure project financing. Governments and regional networks have a

Figure 2.4: Market incentives and barriers to technology transfer

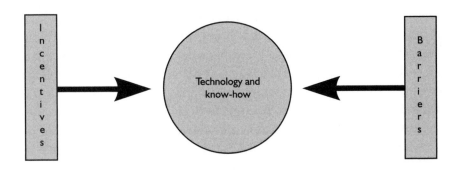

Technology transfer to developing countries

key role to play in facilitating such partnerships by providing up-to-date information. The Promotion of Renewable Energy Sources in South East Asia (PRESSEA) network coordinated by the ASEAN Centre for Energy (ACE), based in Indonesia, is an example of such a network. National representatives for each country involved in the network collect information on such topics as the resource and market potential for renewable energy development in each country involved in the network; the organizations involved with renewable energy in the region; and national plans, regulations and legislation relevant to the development of renewable energy systems. The information is regularly updated and can be accessed on the PRESSEA website.[19] This type of information is needed to raise awareness among potential actors of the possibilities for renewable energy and to encourage them to engage in market development.

The rate of technology transfer is affected by the balance between incentives that encourage the flow of technology (e.g. incentives to encourage investment) and barriers that impede the transfer process (e.g. lack of information and local knowledge). Figure 2.4 shows

---

[19] For further information on the PRESSEA network and details on renewable energy activities and organizations in Southeast Asia, visit the website at <www.ace.or.id/pressea>. See also Chapter 3, Box 3.8.

how barriers and incentives act against each other in the marketplace. The arrow along the bottom indicates that if the central circle representing technology and know-how moves to the right, then technology is being successfully transferred to developing countries. The 'barriers' arrow is working against this movement and the 'incentives' arrow is working with this movement. Incentives are needed to gain interest among actors and momentum in the technology transfer process and to overcome certain barriers. Barriers and the options for overcoming them are covered in more detail in Chapters 4 and 5 below.

It can be seen from Figure 2.3 that both the public and the private sector have an important role to play in the transfer of renewable energy technology. Their roles are outlined in more detail below.

2.2.2 The role of governments

Governments of both developing countries and industrialized countries have a role to play in encouraging and enabling the transfer of renewable energy technology to developing countries.

In general it is the role of government in the recipient developing country to:

- set policies and targets for renewable energy (i.e. short- and long-term renewable energy targets and clear policies on rural electrification);
- raise awareness among different actors regarding the potential market for and applications of renewable energy technologies in that country;
- create incentives which encourage the private sector to take part in the technology transfer process (e.g. tax holidays, reduction or removal of import duty);
- create an environment conducive to participation by private organizations (i.e. institutional structures, supporting infrastructure, regulation and legislation) and where appropriate encourage joint ventures (e.g. by strengthening the legal framework);
- encourage local participation to ensure the requirements of communities are met;
- ensure local adaptation of the technology to protect the environment and resource base.

In general it is the role of government in the industrialized country to:

- provide overseas development assistance in a responsible and sustainable manner;
- encourage the transfer of skills and know-how (e.g. training trainers);
- cooperate with developing country governments regarding awareness raising and dissemination of information (e.g. setting up renewable energy and energy efficiency information networks to exchange information between the two countries or regions; producing a directory of manufacturers and technologies);
- encourage technology R&D and particularly the adaptation of technology to local conditions in developing countries;
- encourage the participation of its domestic private sector in renewable energy technology transfer to developing countries (through e.g. market assessments, trade missions, gear-up export credit agencies to support renewables projects).

2.2.3 The role of international finance institutions

The main role of multilateral and bilateral agencies is in providing grants, soft loans and commercial loans (where risks are perceived to be too high for commercial banks) and investing in feasibility studies, demonstrations, information exchange, capacity building and infrastructure development. Agencies with developmental or environmental goals have a responsibility to make sure that ODA is used to implement projects that will encourage sustainable development and to ensure that all sectors of the community involved share in the benefits from renewable energy. For example, they need to ensure that projects fit in with the sustainable development goals set out in recipient country NSSD and PRSP and encourage the use of local materials where possible. They can also encourage developing countries to make the costs of power generation more transparent; allow independent power producers access to the grid; and reduce subsidies on fossil fuels, thus allowing greater scope for renewables to compete on an equal footing.

2.2.4   The role of private sector organizations

The private sector has an important role to play in providing investment and technology. It also has additional roles:

- as project developer, planning, coordinating and implementing the project (including raising funds or accessing credit for the project);
- in raising awareness of renewable energy technologies and building up confidence in customers (users) and thus stimulating the market;
- in designing and manufacturing technology, engaging in technological R&D, adapting technology to local conditions and market development.

2.2.5   The role of civil society

The main role for civil society is to put pressure on national governments to act in its best interest. It can assist in many aspects of the process, including awareness raising, training and dissemination of information. Small local NGOs working at the grass-roots level can play a key role in providing micro-financing for remote rural areas. They can also help plan schemes and select appropriate systems using participatory planning methods to empower communities. It is important that representative civil society groups make their voice heard at national level.

2.3   Risks and rewards

The rewards and risks likely to accrue to the various actors in the technology transfer process inevitably depend on the roles they play. It is good practice for all companies to assess the rewards and risks of any business transaction before entering into it. The same principles will apply to private-sector organizations considering investment in or transfer of EST to developing countries. Public-sector organizations also need to assess the rewards and risks of encouraging such technology transfer, and local communities and end users need to weigh up the risks and rewards to them. Below is a summary of the main rewards and risks associated with the public- and private-sector actors, and with communities and end users.

2.3.1 Risks

If the real or perceived risks of technology transfer are too great for a particular actor, they will become a deterrent to that actor's involvement and thus create a barrier to renewable energy technology transfer.

Public-sector perspectives The risks which the public sector faces in the process of transferring renewable energy technology are summarized below.

❏ Developing country public-sector risks:

- *Policy/legislation:* New policy and legislation need to support the implementation of renewable energy technology, but there is a risk of damaging competing technology and other sectors in the economy if renewable energy technology is favoured too strongly. The policy and legislation balance must be right to keep a mixture of energy technologies and maintain security of supply.
- *Private-sector competence:* There is a risk that the industrialized country private sector may not be technically competent and may not have technologies that are reliable and durable.
- *Country priorities:* There is a risk that private companies are not committed to fitting in with and assisting the host country's development priorities. Investment by either the private or the public sector of an industrialized country can be used to undermine host country priorities.
- *Duration:* There is a risk that the industrialized country private-sector involvement is not sustained over an adequate period of time for sustainable technology transfer to occur.
- *Intervention:* If not monitored and managed carefully, interventions by government could in fact hinder market development instead of helping it.
- *Dependence on subsidies:* If subsidies are used to encourage the uptake of renewable energy systems, there must be a clear way to

remove these subsidies and leave the market operational. If the market collapses when support is withdrawn, projects will fail and technology transfer will stop.

- *Targeting subsidies:* In general, energy subsidies benefit the rich more than the poor, as the rich consume more energy. It is important to target subsidies so that they reach the desired recipient ('smart subsidies').
- *Willingness to pay:* If renewable energy systems are provided in a demonstration project with heavy subsidies, people are given the impression that they should receive the services very cheaply, if not free of charge. This creates an unwillingness to pay for the services later.
- *Investment risks:* There may be unwillingness to invest due to high initial cost and long payback period. There are also other barriers (planning, institutional and political) to be overcome, so there is no guarantee a project will be implemented successfully. Several years of planning and negotiation may be involved with no guarantee the project will go ahead.
- *Investment in infrastructure, education/training and capacity building:* The public sector needs to invest in supporting infrastructure and capacity building to provide the skilled labour and develop small and medium-sized enterprises to install, operate and maintain the technology properly. If this infrastructure is not in place, the technology will fail.
- *Changing currency exchange rates:* If ODA loans are given in foreign currency, then fluctuations in the exchange rate could increase the value of the loan dramatically in local currency, making it very difficult if not impossible to pay back, unless it is possible to hedge the currency at a fixed rate over a period of time.[20]

❑ Industrialized country public-sector risks:

- Exposing national industries to uncertain markets where return on investments may not be realized and industries may suffer, harming the national economy.

---

[20] If a currency is hedged, the foreign exchange rate is fixed for repayments at an agreed rate over a number of years (for the whole duration of the loan or just part of it).

Private-sector perspectives The risks which the private sector faces in the process of transferring renewable energy technology are summarized below.

❑ Developing country private-sector risks:

- *Technical:* The technology may not be proven for local conditions/ resources, causing projects to fail.
- *Access to finance:* Lack of proven track record or guarantee may make it impossible to obtain a necessary loan.
- *Foreign exchange risk:* Fluctuations in the exchange rate can be disastrous if loans are to be paid in foreign currency and the exchange rate is not hedged
- *Remote locations:* These can cause difficulties in getting access to spare parts or skilled labour, and may cause problems in collecting fees (for reasons of security, honesty and auditability). It is difficult to manage and monitor projects remotely.
- *Changing policy:* This may involve interest rates, tax rates and/or rural electrification plans.

❑ Industrialized country private-sector risks:

- *Planning:* Conflicts of interest may cause planning permission to fail.
- *Technical:* The technology may not be proven for the local conditions/resources, causing projects to fail, giving the technology itself a poor reputation and causing confidence in it to decrease.
- *Resource availability:* Renewable energy resources can be unreliable or intermittent. It can be difficult to secure long-term contracts for biomass fuel supplies. Resource assessments in developing countries are often poor, making it difficult to plan projects effectively.
- *Ability to pay:* The ability of users to pay for the services that the private sector is providing depends on their access to cash or credit systems; the flexibility to pay at times when they gain income from harvesting crops or other seasonal activities; and the option to pay in commodities if they are not able to access markets to exchange crops for cash.

- *Investment risks:* Investors may be reluctant to commit funds due to high initial costs and long payback periods. There are also other barriers (planning, institutional and political) to be overcome, so there is no guarantee a project will be implemented successfully. Several years of planning and negotiation may be involved, with no guarantee the project will go ahead.
- *Financial security:* There is a risk of no return on capital. This could be due to project failure or a cessation of negotiations. For example, the Asian economic crisis of 1997–9 brought negotiation on many projects to a premature end, though more recently several projects are now going ahead again.
- *Ability and willingness to pay:* There is a risk that some rural people cannot or will not pay for the service provided by renewable energy. They may be unfamiliar with a cash economy and/or credit mechanisms.
- *Competitiveness:* There may be financial and institutional barriers that prevent the technology from becoming economically competitive
- *Market risks:* The demand may not be proven in new geographical areas and emerging markets. There is a risk that people may not want the services provided.
- *Confidence:* There is a risk that lending agencies may lose confidence in the market, or that end-users may lose confidence in the technology.
- *Indigenous skills:* A lack of indigenous skilled labour to operate and maintain the equipment properly may cause the system to fail, creating a bad image for the technology.
- *Local skills:* Even when local people are trained, there is often a high turnover of staff as people move away with their new-found skills to urban areas in search of better-paid jobs.
- *Remote locations:* These can cause difficulties in getting access to spare parts and skilled labour, and problems in collecting fees (for reasons of security, honesty and auditability). It is difficult to manage and monitor projects remotely.
- *Quality standards:* The reputation of a manufacturer could be at risk if locally manufactured components are of substandard level.

Standards for systems need to be set carefully as they can restrict what users can afford to buy.

- *Law:* Property rights are a perceived risk to manufacturers. They are reluctant to manufacture components locally for fear of their technology designs being stolen. Another significant concern is the availability and ease of recourse to the law and means for the enforcement of contracts.
- *Changing policy in host country:* This may involve interest rates, tax rates, and rural electrification plans.
- *Political risks:* If a host country is politically unstable, the risks to the private sector investing in the country are higher.

Community/end-user perspectives   A community or end user might see that there is a risk of the technology not being appropriate to their requirements and failing to deliver the services they desire or expect. Managing the expectations of end users and communities is a difficult matter. Sometimes communities or end users are misled about the practical application of the technology by hard marketing, undertaken to boost sales, and the dealer's lack of understanding of the user's requirements. The communities and end users risk not having a voice in selecting the technology most appropriate for them. End users and communities may feel that committing themselves to regular payments over a number of years for the systems is a financial risk they are not willing to take, as it could reduce their capacity to manage their finances during income fluctuations. The community may see new technology as a threat to local cultural, religious or spiritual practices. They may also see it as risk to the social fabric and hierarchical structure of the community, particularly if the elders or leaders of the community do not receive the technology first or have a bigger or better system than others in the village.

2.3.2   Rewards

A reward can be described as what the actor will achieve or gain as a result of taking part in the process of technology transfer. In some cases incentives are put in place to help actors achieve their goals and gain their rewards.

Public-sector perspectives   The rewards which the public sector aim to achieve by transferring renewable energy technology include the following.

❑ Developing country public-sector rewards:

- Reduction of fossil fuel imports.
- Freeing up of indigenous fossil fuels to increase exports.
- Improved trade balance (balance of payments, foreign exchange) as a result of reducing oil imports and increasing foreign investment for local technology manufacture.
- A better energy balance (mix of fuels) which can improve security of supply and reduce dependency on fossil fuels.
- Provision of energy services to rural areas, facilitating rural development.
- Local job creation.
- Local capacity building of skilled labour and institutions.
- Reduction in the cost of previously imported technology.
- Establishment of sustainable businesses locally.
- Environmental protection, e.g. regional, local and indoor air-quality improvements;
- Assisting in poverty alleviation via the provision of energy services to help develop health centres, schools, rural enterprises and communications.

❑ Industrial country public-sector rewards:

- Contribution towards sustainable development which will help secure a better and more stable future for the global economy and thus their national economy.
- Meeting political goals to contribute towards sustainable development.
- Expanding market opportunities abroad to increase exports and improve the balance of trade and economic development of the country.
- Meeting emissions reduction targets at least cost.

Private-sector perspectives   The rewards which the private sector aims to achieve by transferring renewable energy technology include the following.

❑ Developing country private-sector rewards:

- Access to investment capital.
- Access to markets through partnerships and networks.
- Potential to manufacture technology locally, creating local jobs.
- Development of export markets and receiving foreign currency.
- Development of business management, marketing and technical skills, and quality standards in collaboration with foreign partners.
- Improved capability to produce international-level technology.
- Development of local niche markets for renewable energy technology.
- The benefit of combining energy generation with environmental clean-up, e.g. burning biomass residues from industry that would otherwise have to be disposed of at a cost, or biogas production from animal manure.

❑ Industrialized country private-sector rewards:

- Return on investment.
- Market opportunities: potential to expand business via market development for renewable energy technology and/or related markets.
- Increased volumes of manufacture, bringing economies of scale and competitively priced renewable energy systems.
- Potential to manufacture technology locally, bringing down costs.
- Diversification of market base and the development of a second stream of income.
- Access to markets through partnerships and networks.
- Development of niche markets for renewable energy technology.
- Development of local knowledge in collaboration with local partners.
- The attraction of being seen to be a 'green' company or a company that contributes to sustainable development in developing countries, which may increase investment in the company or demand for its products.

Community/end-user perspectives   The rewards that communities and
end users might aim to achieve by adopting renewable energy technology are wide-ranging. They might include

- improved health;
- better health care and educational services;
- access to information and communications technology;
- empowerment of women through reduced drudgery and more time
  for education;
- development of micro-enterprises and jobs;
- improved agricultural production via irrigation;
- extending the day with lighting for social and productive activities;
- improved safety via street lighting and lighting for shops and small
  businesses.

If technology transfer is to take place successfully, there needs to be
a balance between rewards and risks for all actors. In some cases
risk-sharing among actors needs to be encouraged where there is a
disproportionate amount of risk bearing on one potential actor or set
of actors which is preventing them from taking part in the process
and thus creating a barrier to technology transfer.

Chapter 3

Investment in Technology Transfer

3.1 Key investors for renewable energy

The level of future deployment of renewable energy technologies in developing countries is critically dependent on actors being willing to invest in renewable energy technologies and the level of that investment. As with other actors involved in the technology transfer process, those providing or facilitating investment for a particular project or programme will vary depending on the scale and type of project or programme, local policies, their own investment policies, the regulatory and institutional situation, and the way in which the transfer process is being carried out. The range of actors which may invest includes international institutions (e.g. multilateral and bilateral agencies); national institutions (e.g. governments, utilities, academic institutions and banks); private companies and NGOs, which may be national, international or multinational. As the modern understanding of technology transfer includes both hard technology and softer knowledge and skills, investment in R&D and technical transfer needs to be complemented by investment in training, capacity building, information collection and dissemination, institutional structures, planning and policies.

3.1.1 International institutions

International institutions, both multilateral and bilateral, not only invest in the technology transfer process but have a role to play in encouraging the policy and institutional reforms needed to attract other investment (particularly from the private sector). Investment from international institutions is an important source of money for project and programme funding, local training, infrastructure development, information transfer and technical assistance (including feasibility studies and post-project appraisals).

Multilateral organizations   There are several multilateral organizations that have sustainable development or environmental objectives and the capacity to invest in renewable energy projects. Some are already involved to varying degrees in renewable energy investments.

❑ The World Bank: The World Bank Group was founded in 1944 and is the world's largest source of development assistance. It provided more than US$15 billion in loans to its client countries in the financial year 2000. It is owned by 183 member countries, which are represented on the board of directors. Member countries are stakeholders with decision-making power, and thus can influence the activities of the World Bank. Its mission is to fight poverty for lasting results and to help people help themselves and their environment by providing resources, sharing knowledge, building capacity, and forging partnerships in the public and private sectors.[1] It is split into five institutions that work collaboratively towards the overarching goal of poverty reduction:

1 *The International Bank for Reconstruction and Development* (IBRD) provides market-based loans and development assistance to middle-income countries and creditworthy poorer countries to help reduce poverty. Loans typically have a five-year grace period before the first repayment is made and must then be repaid over 15–20 years.
2 *The International Development Association* (IDA) plays a key role in supporting the Bank's poverty reduction mission. Its assistance focuses on the poorest countries that can not afford to borrow from the IBRD. It provides interest-free loans (called credits), technical assistance and policy advice to countries with a per capita annual income of less than US$885. Credits typically have a 10-year grace period and must be paid back over 35–40 years. Borrowers pay a fee of less than 1% of the credit to cover administrative costs. Both IBRD loans and IDA credits are used to support projects and programmes that meet priority economic needs such as health, education, rural development and basic infrastructure. They also help support governments in reforming the structural and social

---

[1] For more information on the World Bank Group see <www.worldbank.org>.

policies that are needed for effective private and public sector development.

3 *The International Finance Corporation* (IFC) was set up to promote growth in developing countries by financing private-sector investment, mobilizing capital in international markets, and promoting technical assistance and advice to governments and businesses. In partnership with private investors, it provides loan and equity finance for business ventures in developing countries. It is profit-orientated, and catalyses private financing by demonstrating the profitability of investment in these countries. The IFC's involvement in projects has given private investors the information, knowledge and confidence to invest greater amounts, across a broader range of companies and sectors, than would otherwise be the case. IFC projects demonstrate that good business opportunities exist in developing countries.

4 *The Multilateral Investment Guarantee Agency* (MIGA) helps encourage foreign investment by providing guarantees to foreign investors against loss caused by non-commercial risks in developing countries. It also provides capacity-building and advisory services to help countries attract foreign direct investment.

5 *The International Centre for Settlement and Investment Disputes* (ICSID) provides facilities for settlement by conciliation or arbitration of investment disputes between foreign investor and host countries.

World Bank lending for renewable energy projects in developing countries accelerated during the 1990s as a confluence of developmental, environmental and social factors began to convince the World Bank and its client countries that renewable energy projects were viable investments.[2] It is interesting to note that no World Bank renewable energy projects were approved by the Bank's board in the 1990s without a Global Environment Facility (GEF) grant also being part of the project.

❑ The Global Environment Facility (GEF): The GEF is the main multilateral funder of renewable energy in developing countries. It is a grant-making vehicle for sustainable development created in the af-

---

[2] Martinot (2001).

termath of the 1992 Earth Summit in Rio de Janeiro. Governed by a 32-member council of donor and recipient countries, the GEF funds projects that are implemented by the United Nations Development Programme (UNDP), the United Nations Environment Programme (UNEP) and the World Bank Group. In the GEF's first decade (1991–2000) it approved $570 million in grants for 48 renewable energy projects in 47 developing and transitional countries. Total project costs have exceeded $3 billion, because GEF grants have also leveraged US$2.5 billion of financing and other resources from governments, other donor agencies, regional development banks, implementing agencies and the private sector.[3]

The GEF's renewable energy projects fall into two categories: removing barriers to markets for commercial or near-commercial technologies; and reducing long-term technology costs through research, demonstration and commercialization. Along with its three implementing agencies (UNDP, UNEP and the World Bank), the GEF seeks to involve and support the private sector and promote commercial and sustainable markets for a variety of renewable applications.

The GEF's overarching goal is to develop sustainable private markets through which to expand the use of renewable energy in developing countries and maximize the social, economic, and environmental benefits this can bring. GEF renewable energy projects involve private firms as manufacturers and dealers, local project developers, financial intermediaries, recipients of technical assistance, technology suppliers and contractors, and project executors. Private project developers, for example, receive financing and technical assistance, while also benefiting from improved regulatory frameworks. Some projects, such as a World Bank/GEF project in Sri Lanka, facilitate innovative micro-financing approaches through local organizations that increase affordability and expand local markets. In addition, a variety of private-sector financing vehicles are emerging, including several through the International Finance Corporation such as the PV Market Transformation Initiative (PVMTI), the Global Renewable Energy and Energy Efficiency Fund for Emerging Markets (REEF), the Solar Development Group (SDG) and the Small and Medium-Scale Enterprise Programme.

---

[3] GEF (2001).

(For information on PVMTI, see below; for details of the other vehicles mentioned here, see Section 3.3.1.) Alongside renewable energy projects, GEF promotes and funds energy efficiency projects; together, the two strands contribute to the GEF climate change mitigation strategies.[4]

The GEF has recently begun to consider long-term project proposals,[5] such as a new 10-year project in Uganda to remove market barriers to private-sector development of about 70 MW of biomass, hydro and solar energy systems. The project will build on a newly enacted private power law through comprehensive capacity building, institution strengthening, and the introduction of regulatory mechanisms facilitating environmentally sustainable private-sector delivery mechanisms.

❏ The United Nations: There are agencies of the United Nations that have some involvement in renewable energy (mainly PV). They tend not to have renewable energy programmes, but provide funding for energy services which help achieve their main aims of improving education and health. These agencies include:

- the *World Health Organization* (WHO), which has funded projects to supply vaccine refrigerators along with technical assistance to health clinics in developing countries as part of its vaccine programme, and has developed standard performance specification and test procedures for vaccine refrigerators. It has also funded various studies into the impact on the health of women and children of emissions and smoke from biomass burnt in open fires for cooking and heating in developing countries;
- the *United Nations Children's Fund* (UNICEF), which has purchased PV systems to power equipment for health centres in remote areas including vaccine refrigerators, lighting systems, water pumps and radios;
- the *United Nations Industrial Development Organization* (UNIDO), which does not fund projects, but has facilitated a number of renewable energy country studies and helped establish centres of

---

[4] Martinot and McDoom (2000).
[5] For more details, see p. 82 on the GEF–World Bank Strategic Partnership.

excellence for rural energy services such as the Centre for Application of Solar Energy (CASE) in Perth, Western Australia;

*   the *United Nations Educational, Scientific and Cultural Organization* (UNESCO), which coordinates the World Solar Summit process that began in Paris in 1993. The profile and importance of the World Solar Summit were raised in Harare in 1996 when heads of state signed up to the Harare Declaration which supported a Solar Decade (1996–2005) and identified 300 solar projects for implementation programmes.[6]

❏ Regional development banks: All the regional development banks have, one way or another, funded projects with a renewable energy element and are well placed to support significant renewable energy projects. The Asian Development Bank (ADB) in particular is looking to fund PV and biomass projects in developing countries. However, although renewable energy and energy efficiency often appear in the energy policies of the regional development banks, the banks lack an integrated strategy to encourage the inclusion of renewable energy in their overall lending programmes or within particular sectors such as health and education.[7]

Multilateral programmes and initiatives   The multilateral agencies such as the World Bank, GEF, UNDP and UNEP are involved in a number of programmes and initiatives which cover renewable energy. These include the Energy Sector Management Assistance Programme (ESMAP), Financing Energy Services for Small-Scale Energy Users (FINESSE), Asia Alternative Energy Programme (ASTAE), PV Market Transformation Initiative (PVMTI), Solar Development Group (SDG), African Energy Enterprise Development (AREED) Initiative[8] and GEF–World Bank Strategic Partnership.

❏ Energy Sector Management Assistance Programme (ESMAP):[9] The overall goal of this donor-funded programme is to increase the avail-

---

[6] Berkovski (1995).
[7] Gregory et al. (1997).
[8] For more details on SDG and AREED see Section 3.3.1.
[9] Adapted from information provided by Charles Feinstein, ESMAP, World Bank. For more information see <www.esmap.org>.

ability of energy services for poverty alleviation and economic and social development. The areas of strategic focus are increasing access to energy services; providing efficient energy services through the development of energy markets; and ensuring environmentally sustainable energy services. ESMAP is managed by the World Bank but also includes support from UNDP. A key function of the ESMAP programme is dissemination of the knowledge generated through projects.

ESMAP's principal objective under the renewable energy theme is to get these technologies into the mainstream agendas of local governments and development institutions in order to contribute to international efforts to provide clean energy. ESMAP's approach includes regional or country pre-investment work, country-specific project identification, technical assistance and, in some circumstances, the introduction and commercial demonstration of new, non-conventional or hybrid energy sources with potential for promising applications in rural or peri-urban poor areas.

❏ Financing Energy Services for Small-Scale Energy Users (FINESSE):[10] FINESSE was launched in 1989 by the World Bank, the US Department of Energy and the government of the Netherlands, with the aim of encouraging market implementation of renewable energy technologies. Coverage was originally limited to Asia (with participating countries including Indonesia, Malaysia, the Philippines and Thailand). In 1994 a programme for South America was established, followed in 1996 by a programme for southern Africa, with involvement from UNDP and the Southern African Development Community (SADC). FINESSE is now predominantly coordinated by UNDP. There is a successful UNDP FINESSE activity in the Philippines, run in cooperation with the Development Bank of the Philippines (DBP), which includes training of DBP staff in renewable energy project appraisal methodology and risk assessment. FINESSE aims to work with participating countries to facilitate the identification of potential renewable energy resources and to assist them in establishing priorities to use those resources. To achieve this, FINESSE channels grants and soft loans

---

[10] Gregory et al. (1997).

through a range of lending intermediaries (including private companies, utilities, NGOs and commercial finance institutions) to support the financing of small-scale energy users' projects. FINESSE assists regional development banks and other investors in understanding the investment risk and benefits for renewable energy projects, as conventional methods of project financing are designed for large-scale fossil-fuel-based power plants, which have a very different investment profile from small-scale renewable energy projects. Thus its activities include raising awareness of and training in project appraisal methodology suitable for financing small-scale renewable energy projects.

❑ Asia Alternative Energy Programme (ASTAE):[11] The Asia Alternative Energy Programme (ASTAE) was established by the World Bank in 1992 with support from the GEF, in part to help implement FINESSE in Asia. The goal of ASTAE was to bring sustainable energy into the mainstream of Asian development by 'greening' the World Bank lending to the power sector in this region. The programme has been successful in increasing the share of alternative energy in its Asian power sector loan portfolio, which in the financial year 1999 reached 46.3%. As of June 2000, 38 projects were either in the pipeline, approved or completed, and it is projected that the implementation of these projects will replace around 1 GW of conventional capacity.

ASTAE performs the following activities: identification, preparation, appraisal and supervision of renewable energy and energy efficiency investments, supported by the World Bank and by the GEF; economic and energy sector work, including formulation of policies to promote environmentally sustainable renewable energy and energy efficiency options; technical assistance and capacity building; training of World Bank staff and borrowers; coordination with donor agencies; and mobilization of resources for alternative energy development. ASTAE's focus on alternative energy project development has been complemented and supported by awareness creation and capacity-building efforts within the World Bank and in its Asian recipient countries.

---

[11] Adapted from information provided to the G8 Renewable Energy Task Force by Enno Heijndermans, World Bank.

The factors that have contributed to the success of ASTAE are a sectoral/regional focus; a results-orientated approach with clear mandates and lending targets; mobilization and strategic use of funds, where ASTAE funds are used for preparatory work and funds from the World Bank and other sources, both public and private, are used for subsequent work; partnerships both within the Bank, between ASTAE and other programmes, and outside the Bank, between ASTAE and stakeholders in governments, the private sector, NGOs, multilateral institutions and research institutes; long-term donor support; and timely implementation. ASTAE has successfully tapped into the client countries' growing energy demands, local private-sector interest, the World Bank's increased emphasis on poverty alleviation and environmental protection, and the commitment of bilateral and multilateral donors.

The achievements of ASTAE in the development of the alternative energy-lending portfolio have been significant. However, if it is to have a wider impact in poverty alleviation and reduction of environmental damage in Asia there needs to be a shift from the project-by-project approach to programmes that create an enabling environment for alternative energy project implementation. This should include the development of favourable policy and regulatory frameworks and effective institutional structures, and the building of broad-based local capacity to develop and implement alternative energy projects and programmes.

❑ PV Market Transformation Initiative (PVMTI):[12] PVMTI was launched by the International Finance Corporation (IFC) as an innovative investment facility designed to provide finance in private-sector projects to encourage the development of markets for PV. The principal aim of PVMTI is to accelerate the sustainable commercialization and financial viability of PV technology in the developing world. PVMTI aims to address market barriers by making available appropriate finance and stimulating business activity. The specific focus is to stimulate PV business activities in India, Kenya and Morocco. This is achieved through:

---

[12] Based on information provided by IT Power, UK.

- providing finance for sustainable and replicable commercial PV business models, according to individual business plans submitted through a competitive bidding process;
- financing business plans with commercial loans at below-market terms or with partial guarantees or equity instruments;
- providing technical assistance to PV businesses on planning, financing operations and technology.

The duration of the PVMTI programme is 10 years, from 1 July 1998. At the end of 1999 PVMTI funds totalled US$120 million, US$30 million of which was from the GEF and the remaining US$90 million raised by the IFC from outside investors.[13] The first investment was approved in September 1999 and by June 2001 more than 70% of available funds had been allocated (or referred to the IFC for approval) for a wide variety of PV investments. The projects funded under PVMTI are selected on a competitive tender basis. There is a minimum and maximum size for PVMTI investment in a single entity; these limits vary from country to country. To ensure adequate commitment from proposers, a minimum level of cofinancing is required before a PVMTI investment can be authorized.

At the outset a market analysis of the three chosen countries was undertaken, with an evaluation of the PVMTI business case and a summary of the potential PVMTI investment flows. A key theme underpinning the potential success of PVMTI is the emphasis on the unique situations of each country's business environment and the acknowledgment that one solution is not applicable to all.

❑ GEF–World Bank strategic partnership:[14] The principles of the strategic partnership include: targeted increases in GEF resources, with a proposed interim target of $150 million annually; a long-term country-based business planning approach (i.e. five- to ten-year development plans); and a simplified approval process. The strategic partnership

---

[13] Martinot (2001)

[14] Adapted from information provided to the G8 Renewable Energy Task Force by Ted Kennedy, World Bank. For more information on the strategic partnership see <www.gefweb.org>.

was set up to expand and increase the effectiveness of the renewable energy activities of the World Bank and GEF and to shift efforts from an individual project approach on to long-term, programmatic pathways. In this way, it will provide developing countries with the time and resources required to develop renewable energy markets and technologies in a comprehensive and sustainable way.

The first project submitted under the partnership is 'energy for rural transformation in Uganda'. This project will provide resources to remove market barriers for the development of around 70 MW of biomass, hydro and solar renewable energy capacity, concentrated in the private sector. It is anticipated that the development of rural energy sources will be significantly accelerated in line with rural development needs and objectives, and rural electricity connections will be increased from 1% to 10% over the life of the project. This will provide a significant shift away from diesel power sources. Investments will build on new energy supply opportunities created by a recently enacted private power law. Activities will be directed at capacity building, institutional strengthening and the introduction of 'light regulation' approaches designed to facilitate growth in environmentally sustainable private-sector delivery mechanisms. The loans are to be divided into three tranches linked to the accomplishment of key objectives, including the achievement of specific infrastructure changes, finalization of long-term renewable energy plans and targets for renewable energy system installations. It is anticipated that the long-term commitment to the development of renewables in the country, with stated targets backed up by funding, will provide the stability and confidence necessary for private-sector market development.

Bilateral agencies   Various bilateral agencies are involved with funding renewable energy projects. One of the first large PV projects, the Regional Solar Programme in the Sahel, was funded by the European Development Fund (EDF). The programme installed 626 PV pumping systems and 644 community systems in nine Sahelian countries between 1991 and 1997. The European Commission funds renewable energy projects in Europe and developing countries from budgets controlled by different Directorates-General (DGs). For example, the

---

Box 3.1: Solar home systems in the island state of Kiribati, South Pacific

The European Commission funded a pilot installation of 250 photovoltaic solar home systems in the island state of Kiribati in the South Pacific. The systems were in place by 1995 and were part of the 1988 PV follow-up programme of the Pacific Regional Energy Programme (PREP). The solar energy company (SEC) in Kiribati installed and maintained the systems, which remain the property of the SEC; households pay a fee of around A$10 each month for the service provided. The SEC maintains the systems on a regular basis and replaces components when they come to the end of their life, including the PV panels. In 1998–9 an evaluation of the pilot project was carried out. The evaluation looked at the benefits the systems were providing to the islanders, and the technical, environmental and financial sustainability of the systems.

The results of the analysis showed that the PV solar home systems were providing benefits to the users in the form of better-quality light for cooking, eating, reading, weaving, socializing and preparing to go fishing at night. The lighting also provided better security and safety. The solar home systems also powered radio/cassette machines and CB radios for communication between islands. The analysis showed that the solar home systems were better for the local environment than transporting kerosene and diesel to the outer islands (due to risks of spillage from barrels and damage to the local marine environment). The systems were found to be working well, especially as some parts of the system were assembled locally (the controllers and lighting ballasts) so that local technicians were able to service and maintain the equipment relatively easily. It was found that in order for the SEC to become financially sustainable (i.e. cover overhead costs, staff costs and replacement parts), the number of systems on the outer islands needed to be increased to over 1,000. Subsequent to the report findings the European Commission DG for Development has agreed to fund the installation of another 1,000 systems so that the SEC can become financially sustainable.

Source: Gillett and Wilkins (1999).

Directorate-General for Development is funding the installation of an additional 1,000 solar home systems in Kiribati in the South Pacific (see Box 3.1). The Directorate-General for Energy (before it was merged with transport to become DGTREN) funded the setting up of two renewable energy information networks, one in southern Africa (called SAREIN) and the other in Southeast Asia (called PRESSEA; see Box 3.2 for information). It also supported South Africa in funding PV systems in 1,000 schools. The earliest major project funded was the installation of 850 PV systems for health centres in Zaire in 1984.

Other bilateral agencies are active in funding many other renewable energy projects, too numerous to mention here. The more active agencies include:

Box 3.2: Promotion of renewable energy sources in Southeast Asia (PRESSEA)

A significant barrier to the development of renewable energy markets in developing and rapidly emerging economies has arisen from the difficulty of getting reliable market information. Without reliable, easily accessible information an investor will perceive a market to be high-risk. Following a feasibility study, the European Commission's DGTREN supported the launch and operation of the PRESSEA renewable energy information network in Southeast Asia. The network built on the existing national and intergovernmental renewable energy structures in the region. Regional coordination was provided by the ASEAN Centre for Energy (ACE); the national focal points for Indonesia, Malaysia, Philippines, Thailand and Vietnam were the representatives of the ASEAN intergovernmental subsector network on new and renewable sources of energy (NRSE SSN).

The primary achievements of the project have been:

- freely available market information – a regularly updated website containing the equivalent of more than 700 pages of market and contact information, much of it from official sources; and
- sustainability – the project has been integrated into the ASEAN five-year plan of energy cooperation, and the core PRESSEA activities will continue after the project has finished under the authority granted to ACE through this plan.

This latter achievement is significant in that the project has facilitated the operation of the ASEAN NRSE SSN. The successes of the project are attributed to:

- a strategy of adding value to the existing NRSE infrastructure in ASEAN;
- a strong local presence though the offices of ACE and the European collaborators; and
- delegation of market research and website development to the local collaborators, rather than attempting to undertake the work in Europe.

Source: Slesenger et al. (2001). For more information see <www.ace.id.or/pressea>.

- the German Agency for Technical Cooperation (GTZ);
- the Danish Agency for International Cooperation (DANIDA);
- the Agency for Development and Cooperation in the Netherlands (DGIS);
- the Swedish International Development Cooperation Agency (SIDA);
- the United States Agency for International Development (USAID);
- the Canadian International Development Agency (CIDA).

3.1.2   National institutions

National organizations that invest in renewable energy include government departments and agencies, energy utilities, academic institutions and national banks.

Government   Industrialized country governments can invest in renewable energy projects in developing countries via their international aid agencies, through relevant government departments (e.g. environment and trade), through their export promotion agencies and through research and development agencies. These investments help raise private-sector awareness of best practice for renewable energy technology application and of lessons learned in developing countries. They also raise awareness of market opportunities and help to establish business links and partnerships between their country's industrial firms and developing country partners.

Developing country governments can invest in pilot and demonstration projects in their own countries, but can also invest time and money in various other activities that will encourage the implementation of renewable energy systems and attract investment from other organizations, nationally or internationally. Some of these activities have a direct impact and others are more indirect. Activities that can have a direct impact on encouraging investment include reducing import tax on renewable energy technology components; levelling the playing field for renewable energy and fossil fuels, which often benefit from both hidden and direct subsidies; providing loan guarantees; and working with international lending institutions to establish credit channels necessary to provide household-level loans. Government activities that can have an indirect impact on investment in renewables include adoption and enforcement of technology quality standards; training programmes; provision of access to and dissemination of information; setting targets and establishing programmes for renewable energy development; planning regulations; and research and development to adapt technologies to local conditions. It is important for both donor and recipient country governments to coordinate their respective efforts in order to achieve the greatest benefit from potential investments.

One way of encouraging investment from the private sector is to allow independent power producers to sell electricity to the grid (via power purchase agreements), and to set a target for renewable energy generation. This is being done in several developing countries which are going through the process of liberalization. Franchises can also be given to private companies to provide power to remote villages which are not easily accessible by the grid. In such cases renewable energy systems can often provide the most practical and economic energy solution in remote areas.

Utilities   Power utilities in developing countries (often government-owned, although recent trends have been towards privatization and liberalization) can either invest directly in renewable energy technologies or encourage others to do so through access to the grid via power purchase agreements, as described above. Direct investment in renewable energy projects has up until now been mainly in large-scale hydro and geothermal.

Academic institutions   Academic institutions contribute indirectly to investment in renewable energy through various activities, including technology research, development and adaptation; developing standards; training; information collection, analysis and dissemination; working with NGOs and other actors; and managing government-funded programmes.

National banks   National banks in developing countries, particularly those focused on agriculture and rural development, are geared up to providing access to credit in rural areas and have branches suitably located for communities needing to access funds for renewable energy projects. Some national banks in developing countries are beginning to consider investing in renewable energy projects. For example, the Vietnam Bank for Agriculture and Rural Development (VBARD) is involved in the financing for solar home systems in Vietnam (see Case Study 3 in Annex 1), and the Agricultural Bank in Zimbabwe is also involved in financing for solar home systems.

3.1.3 Private companies

Private companies may be local, national, international or multinational. The major role for manufacturers is investment in the research, development, manufacture and supply of renewable energy technologies. Just as important are the installation, maintenance and after-sales service of technologies, which in many cases are carried out by a company other than the manufacturer. There is a growing trend towards private investment in the energy field. With increasing concern over climate change, large insurance companies are beginning to realize that they are in the front line to lose out financially if climate-induced natural disasters continue to worsen. This, along with the competitive advantage of being 'seen to be green', is encouraging a growing number of private companies to invest in cleaner and more sustainable forms of energy. Innovative financing mechanisms are beginning to be developed which attract and facilitate investment (see Section 3.3.1 below). With such mechanisms in place, it is hoped that the trend towards private investment for clean, sustainable energy will continue.

Private banks in developing countries are also starting to consider financing renewable energy projects; for example, commercial banks are involved in providing credit for wind power development in India, facilitated by the Indian Renewable Energy Development Association (IREDA). Commercial banks are also involved in providing credit for solar home systems in Sri Lanka (extending World Bank credit for solar home system dealers) and in Kenya.[15]

3.1.4 Non-governmental organizations

There are many local, national and international NGOs, foundations and industry associations supporting the transfer of technology to developing countries. Activities undertaken aim to develop capacity, demonstrate technology, disseminate information and raise awareness. NGOs range from small not-for-profit organizations to industry consortia involving utilities and technology manufacturers and suppliers. There are a small number of NGOs that are either successfully

---

[15] Personal communication with Eric Martinot (GEF).

Box 3.3: Ensuring access to micro-credit via existing institutional structures

In 1998, the French NGO Fondation Energies pour le Monde (FONDEM), together with the Ministry of Health and the Directorate for Energy, launched a rural electrification project in Madagascar. The project included three components:

1 social (electrification of rural health centres);
2 domestic (electrification of households through micro-credit); and
3 economic (electrification for productive applications through fee-for-service arrangements).

The objectives of the project are to:

- electrify approximately 50 rural health centres through PV systems and set up a sustainable scheme of operation and maintenance;
- promote the use of PV for surrounding households within a province of Madagascar through appropriate financial and distribution mechanisms (micro-credit scheme with a local bank and sales, erection, maintenance through local retailers); and
- supply electricity for the economic development needs of entrepreneurs in the area.

Specific attention is paid to the awareness and involvement of the various local stakeholders (e.g. institutions, banks, retailers, suppliers and users). For example, the project aims to raise awareness of how renewable energy can meet local demand; of the assets and limitations of renewable energy systems; and of information concerning operational costs. To address barriers, two things have been done. For the systems powering health centres, a scheme for maintenance and provision for replacement of components involving the users, the Ministry of Health and UNICEF has been set up; and for PV systems powering households and workshops, the micro-credit scheme has been adapted for the duration of the credit.

Lessons that have been learned indicate that there is a need for complementary interests between the stakeholders (government, bank, distributors, end users); a range of systems to match the various needs and financial capacities; and strong awareness raising among the various stakeholders.

Source: Adapted from information provided by FONDEM, France.

---

providing direct financing for renewable energy projects at the grassroots level in developing countries or involved with the facilitation of capital flows from international markets to local borrowers. An example of the former is Grameen Shakti in Bangladesh, a non-profit dealer selling and serving solar home systems, which is a prominent investor, providing credit for SHS in Bangladesh.[16] The Solar Electric Light Fund (SELF) and Enersol Associates are among the largest and best-known NGOs of the latter type. Not many NGOs have been successful in making the

---

[16] Martinot et al. (2000).

transition from technical and development assistance to financial services, though there are many NGOs active in projects that include
renewable energy. For example, the French NGO Fondation Energies
pour le Monde (FONDEM) has been the catalyst for the installation of
renewable energy systems in several different countries including Vietnam, China, Morocco and Madagascar.

If projects are to be successful it is important to make sure that micro-credit facilities are available to rural communities (see Box 3.3).
Some development NGOs and aid agencies, for example, the British
Red Cross and Médecins Sans Frontières, include renewable energy
technologies to power technology supplied for health clinics, schools,
and clean water and sanitation systems.

3.2 Investment trends

It is difficult accurately to measure the amount of investment going
into energy technology transfer, first because investment is being made
both into the technology itself and into training, knowledge transfer
and supporting infrastructure, and second because of the large range
of different actors investing in technology transfer, which makes it
even more complicated to monitor investments. It is reasonable to assume that there is a relationship between international financial flows
and international technology transfer. Unfortunately, statistics are not
collected and analysed in a way that makes it easy to determine the
exact relationship. However, a rough way of estimating the level of
investment going into energy technology transfer is to look at aggre-
gate data on international financial flows, even though they obscure
some important variations.[17]

Capital investment is crucial for energy development. Difficulties in
attracting capital for energy investment can slow economic development. During the 1990s global energy investment averaged just over
1% of GDP (ranging from US$240 billion to US$280 billion a year). It
was concluded in the recent *World Energy Assessment* that capital
market size does not appear to be a limiting factor for energy sector

---

[17] IPCC (2000), ch. 2.

financing today and is not likely to be one in the future.[18] The key is attracting available investment capital to the energy sector and making sure the capital is invested in the most effective way and accessible to those who need it. This will involve capacity building and awareness raising within international finance institutions, which will need to adopt more flexible investment criteria for renewables projects. In addition, many developers are small and may have limited business experience, so will require training to help them package projects in a way that is attractive to investors, creating bankable proposals.

Private-sector investment will be extremely important for future transfer of technology. This is reflected in the value of private flows compared to official development assistance. Foreign direct investment (FDI) totalled around US$240–250 billion in 1997, compared to US$40–50 billion from official development assistance.[19]

There are several types of international financial flows which support technology transfer.[20] These include official development assistance (ODA), official aid (OA), commercial loans, FDI, commercial sales, foreign portfolio equity investment and venture capital. The transfer process for a particular technology may receive investments from one or more sources. There is no direct relationship between specific actors and different types of investment flow: actors do not always use the same route for investment.[21]

3.2.1 Official development assistance (ODA) and official aid (OA)

ODA includes grants and interest-free or subsidized loans to developing countries. OA covers funding of similar types but directed to EIT. The funding mainly comes as bilateral aid directly from OECD countries or indirectly from government contributions through multilateral agencies. ODA remains the main method for government support to developing countries, particularly the least developed countries, but includes financial flows other than technology transfer investments.

---

[18] *WEA* (2000), ch. 9.
[19] Grubb et al. (1999).
[20] WBCSD (1998).
[21] IPCC (2000), ch. 2.

The OECD's Development Assistance Committee (DAC) collects statistics on ODA flows. Table 3.1 shows bilateral ODA commitments to energy from OECD countries from 1989 to 1999. It can be seen that the total commitments peaked in 1995 at around US$5.46 billion and that there was a drop in commitments from US$4.2 billion to US$1.9 billion during the Asian economic crisis from 1997 to 1999, reflecting the slowdown in economic growth and demand for energy during these years. The OECD's Creditor Reporting System (CRS) is an information system containing more detailed information on the financial flows. CRS data in Table 3.2 show bilateral ODA commitments for renewable energy from 1989 to 1999. Again renewable energy commitments are shown peaking in 1995 at US$1.9 billion and the Asian economic crisis is reflected in a reduction from US$1.1 billion to US$0.4 billion between 1997 and 1999. Large hydro accounted for around 75% of renewable energy commitments from 1993 to 1998 but has started to decrease since, in part being replaced by investments in geothermal, solar and wind since 1996.[22] It has been estimated that only around US$3 billion of the ODA for renewables between 1980 and 2000 was for sources other than large hydro.[23]

Japan committed by far the greatest amount of bilateral ODA to energy between 1989 and 1999, accounting for around 54% of total commitments. Other countries with significant contributions include Germany (12% of total commitments over this period), France (6.4%), the United States (6.1%), the United Kingdom (4.7%) and Italy (4.3%).

As for ODA energy commitments to renewable energy, Japan has also provided the largest amount over the period 1989–99, accounting for 57% of the total. Other countries with significant contributions over this period are Germany (11% of total ODA commitments to renewable energy), France (8.0%), Italy (6.7%) and Sweden (3.7%). Table 3.3 shows the percentage of ODA energy commitments for renewable energy in each year from 1989 to 1999. It can be seen that renewables represented 45% of ODA energy commitments in 1998, but only 21% by 1999.

---

[22] Ibid.

[23] Personal communication with Eric Martinot (GEF).

Table 3.1: Bilateral ODA commitments to energy, 1989–99 (US$m)

OECD country	1989	1990	1991	1992	1993	1994	1995	1996	1997	1998	1999
Australia	7.23	4.35	2.87	17.85	37.38	28.99	22.49	18.62	15.68	10.40	4.05
Austria	20.32	8.67	88.21	77.20	68.40	62.48	12.96	9.39	1.90	2.14	4.30
Belgium	5.71	4.49	13.25	10.86	9.64	9.26	3.52	3.87	2.26	1.15	1.35
Canada	70.63	40.28	61.72	44.32	50.85	24.03	20.39	82.12	55.40	47.05	43.71
Denmark	10.30	23.11	28.67	11.79	38.23	87.81	26.79	56.66	32.10	30.65	19.34
Finland	72.80	15.59	139.21	17.33	19.39	1.60	2.10	30.53	4.18	5.28	9.48
France	342.74	600.08	222.13	234.09	239.16	172.08	155.65	183.40	103.82	208.53	–
Germany	517.31	667.47	505.04	358.13	668.19	227.09	300.66	448.84	506.73	189.38	300.50
Greece	–	–	–	–	–	–	–	–	–	0.13	–
Iceland	–	–	–	0.32	–	0.05	0.79	0.79	–	0.00	–
Italy	243.51	282.31	564.53	237.37	35.93	129.10	116.77	34.61	28.59	0.14	6.63
Japan	495.24	546.95	1959.95	1141.94	1898.44	2565.89	4144.17	2061.27	3084.33	1747.44	1244.53
Luxembourg	–	–	–	1.31	0.98	–	–	–	1.09	1.41	1.56
Netherlands	18.13	37.35	2.21	19.11	24.65	37.63	77.43	68.45	40.60	34.21	15.23
New Zealand	–	–	1.02	0.92	0.97	2.35	2.35	–	–	1.22	1.05
Norway	13.50	45.17	118.28	24.54	62.24	25.15	140.68	76.18	61.55	53.09	36.34
Portugal	–	–	–	0.19	–	1.91	0.09	0.27	1.78	0.55	0.09
Spain	195.90	–	485.55	113.67	53.56	110.38	34.84	79.79	32.23	60.51	5.09
Sweden	14.84	50.06	59.77	69.31	103.92	70.53	55.29	103.56	73.95	31.71	29.45
Switzerland	–	4.09	15.24	1.42	1.10	4.24	2.12	0.10	–	1.24	9.29
United Kingdom	227.62	206.50	486.69	133.11	127.72	125.93	123.28	95.87	77.67	80.82	119.75
USA	282.94	308.32	308.32	293.56	276.22	206.80	213.40	166.00	91.52	135.80	87.56
European Community	–	–	–	–	–	–	28.05	–	–	–	–
Total	2,538.72	2,844.79	5,062.66	2,808.34	3,716.97	3,893.30	5,455.77	3,520.32	4,215.38	2,642.85	1,939.30

Source: Development Assistance Committee of the OECD.

Table 3.2: Bilateral ODA commitments to renewable energy, 1989–99 (US$m)

OECD country	1989	1990	1991	1992	1993	1994	1995	1996	1997	1998	1999
Australia	0.48	–	–	–	0.38	–	0.01	18.44	8.20	1.32	1.64
Austria	–	–	1.63	25.20	21.01	52.27	–	5.44	0.42	0.84	1.47
Belgium	–	–	–	–	2.27	3.52	1.03	0.08	0.27	0.55	0.29
Canada	1.34	4.78	5.85	14.26	21.28	11.17	11.54	36.32	27.53	3.23	30.68
Denmark	–	8.17	5.52	0.34	0.14	33.32	5.37	19.19	20.19	9.29	7.61
Finland	4.84	0.93	41.57	7.93	9.31	–	–	15.24	0.91	0.84	–
France	261.48	239.27	24.57	8.16	33.50	28.45	42.06	42.66	71.37	88.43	23.07
Germany	–	27.29	66.22	112.22	10.28	108.08	134.56	251.55	91.22	190.12	238.54
Italy	214.71	92.41	112.32	82.67	1.14	123.37	74.87	3.02	23.40	0.09	2.60
Japan	283.12	145.14	701.19	121.06	584.94	680.06	1518.77	943.28	816.01	389.23	35.38
Netherlands	0.39	0.37	0.76	5.57	0.94	2.44	2.87	27.96	18.91	19.11	9.26
New Zealand	–	–	–	–	–	–	0.46	–	–	–	–
Norway	0.03	7.91	62.82	6.62	6.19	4.60	87.53	13.58	4.66	9.54	2.36
Portugal	–	–	–	–	–	–	0.30	–	–	–	–
Spain	307.87	–	58.19	37.78	–	2.91	33.64	10.00	0.64	41.07	19.80
Sweden	–	–	39.37	–	0.04	0.92	2.09	11.95	0.07	1.13	9.92
Switzerland	–	0.70	13.11	–	1.01	0.74	3.12	0.59	0.73	–	–
United Kingdom	28.68	44.67	7.94	53.26	43.36	35.93	–	0.39	4.33	3.64	0.09
USA	–	–	–	–	–	–	–	0.97	3.70	7.14	6.89
European Community (EDF)	38.71	10.77	54.39	1.39	2.52	3.86	0.19	24.26	0.32	5.92	20.46
Total	1141.66	582.41	1195.42	476.45	738.29	1091.64	1918.40	1434.92	1092.88	771.51	410.04

Source: Creditor Reporting System of the OECD.

Table 3.3: Renewable energy commitments as proportion of total ODA energy commitments, 1989–99 (US$m)

	1989	1990	1991	1992	1993	1994	1995	1996	1997	1998	1999
Total ODA energy	2,539	2,845	5,063	2,808	3,717	3,893	5,456	3,520	4,215	2,643	1,939
ODA renewables	1,142	582	1,195	476	738	1,092	1,918	1,435	1,093	771	410
Renewables share (%)	45.0	20.5	23.6	17.0	20.0	28.0	35.0	40.8	26.0	29.2	21.2

Source: Development Assistance Committee of the OECD.

ODA is not evenly distributed to recipient countries. Bilateral ODA for energy has been focused on Asia, the region with the fastest-growing energy demand in recent years. Ten Asian countries have received 63% of bilateral ODA for energy since 1980 (India, China, Indonesia, Pakistan, Philippines, Thailand, Vietnam, Malaysia, Sri Lanka and Bangladesh). Although ODA for energy decreased during the Asian economic crisis, the total amount of ODA for all sectors increased, from US$47.6 billion in 1997 to US$51.5 billion in 1998.[24] This trend was in part due to a response from bilateral and multilateral aid donors to the need for short-term measures to help developing countries recover from the crisis.

3.2.2   Commercial loans

Loans at market rates are advanced by international institutions such as multilateral agencies as well as by commercial banks. As multilateral agencies also channel ODA and OA loans, the boundaries between investment flows of different types become blurred. Foreign private capital flows to developing countries rose steeply in the early 1990s until the Asian economic crisis, when they fell sharply: total global international bank lending to developing countries dropped from US$86 billion in 1996 to US$12 billion in 1997 to minus US$65 billion in 1998 (the negative figure means that the amount of loan repayments from developing countries to private investors in other countries was US$65 billion more than the foreign investment being made to developing countries in that year).[25] World Bank Group lending for energy investments fell overall between 1995 and 1999, but the Bank's commitments for energy efficiency and renewable energy increased.[26] In all, 17 renewable energy projects were approved for World Bank funding by the end of 1999. Total funding for these projects came to around US$4 billion, of which around US$230 million came from GEF grants, US$704 million from World Bank loans and US$3.1 billion from outside

---

[24] OECD (1999).
[25] Ibid.
[26] IPCC (2000), ch. 2.

investors.[27] At the end of 1999, 13 additional renewable energy projects had received approval for GEF grants but were waiting for World Bank approval.

3.2.3 Foreign direct investment

FDI involves direct investment in plant or equipment by an organization from another country. FDI flows involve both financial inputs and the capitalization of technology, knowledge, skills and other resources that represent a stock of assets for production.[28] An increase in FDI does not necessarily mean an immediate increase in technology transfer, as the relationship between FDI and technology transfer is complex. Private companies looking to invest via FDI can afford to be very selective in where they invest. FDI is more likely to occur in countries with large market potential, little corruption, local skilled labour, good natural resources and sound social and civil order, as these factors increase the chances of good return on investment.[29] However, FDI has become increasingly important for the introduction of new technologies in developing countries. Some recent analysis shows a clear correlation over a group of 20 developing countries between a decline in energy intensity (primary energy use per unit of GDP) and increase in FDI.[30] The most likely reason for this is that modern technology is increasing in developing countries with an increase in FDI, helping them to leapfrog the old inefficient technology used formerly.

Although the rate of increase slowed during the financial crisis of 1997–9, FDI flows to developing countries remained resilient as they are reasonably long-term in nature. FDI flows to developing countries were estimated to be US$171 billion in 1998, rising to US$192 billion in 1999, representing 2.8% of the GDP of the recipient countries.[31] Between 1995 and 1996 only 34% of FDI flows went to developing

[27] Martinot (2001).
[28] ICC (1998).
[29] WBCSD (1998).
[30] Mielnik and Goldemberg (2001).
[31] World Bank (2000b).

countries,[32] and if this proportion is to be increased, policy, legal and institutional conditions need to be strengthened to attract a greater proportion of investment to developing countries. South–South private investment from newly industrialized countries is an important element, accounting for one-quarter of total private investment in 1992.[33]

3.2.4   Commercial sales

This is the sale and corresponding purchase of equipment or knowledge. Around three-quarters of international technology transfer arises from trade of technology in commercial markets.[34] Imports of capital goods to newly industrialized countries in Asia (the 'Asian Tiger' countries) have consistently represented between 20 and 40% of gross domestic investment,[35] although the percentage for renewable energy technologies is not clear.

3.2.5   Foreign portfolio equity investment and venture capital

Foreign portfolio equity investment (FPEI) and venture capital involve purchase of stock or shares of foreign companies directly or through investment funds. Investments are primarily from the private sector, although other organizations also invest in this way. These methods of investment make money available to companies in developing countries, which can then be used to purchase technical equipment and invest in business management and technical training. Venture capital is longer-term and higher-risk, and involves a greater degree of management control by investors than FPEI. Returns on venture capital are usually taken in capital gains and fees rather than dividends. Venture capital is usually focused on early-stage investment in technology-based companies.

---

[32] IPCC (2000), ch. 2.
[33] UNCTAD (1996).
[34] IPCC (2000), ch. 2, OECD estimates.
[35] UNCTAD (1998).

The capitalization of stock markets in developing countries increased by a factor of ten between 1986 and 1995. Contributing factors to this increase in capital investment were the liberalization of stock markets in developing countries, the globalization of financial markets and the concentration of financial resources in institutional investors (e.g. pension funds and insurance companies).[36] FPEI investments are shorter-term than FDI so generally fluctuate more rapidly. During the Asian economic crisis, FPEI focused on industrialized countries as returns on capital were better than in developing countries. This was reflected by a drop in equity investments in developing countries from US$38 billion in 1997 to US$12 billion in 1998.[37] Venture capital tends mainly to be invested in small, technology-based companies in industrialized countries. In 1998–2000 there was a flurry of venture capital investments in internet-based 'dot.com' companies with large potential to make profit, but with no actual profits on their books for the first few years. This bubble burst in late 2000, with shares dropping sharply in these companies and venture capital investors having to look elsewhere for profitable investments.

3.2.6 Other financial flows

These include export credit agencies (ECAs), activities supported by NGOs, education, training and related investments that do not fall into any of the categories outlined above.

Governments often encourage export of domestic technology through export promotion programmes, trade missions and market intermediation, and through ECAs offering export credits, project financing, risk guarantees and insurance on more favourable terms than commercial banks or insurance houses. The participation of ECAs attracts significant private capital investments to projects.[38] Over 80% of ECA financing to developing countries comes from G7 countries.[39] ECA

---

[36] IPCC (2000) ch. 2.
[37] OECD (1999).
[38] Maurer and Bhandari (2000).
[39] The G7 countries are: Canada, France, Germany, Italy, Japan, United Kingdom and the United States of America.

flows in the mid-1990s were on average around US$110–120 billion annually, over twice the value of ODA to developing countries (US$50 billion) and three times that of concessional financing from multilateral development banks (US$40 billion).[40] China and Indonesia accounted for more than half of the total ECA flows in this period.[41] However, ECAs in general are not geared up to support exports of renewable energy technology as they are unfamiliar with the risks involved.

Grants from NGOs to developing countries have remained in the range US$5–6 billion per year since 1990. It is not clear how much of this is contributing towards renewable energy technology transfer. Some NGOs also assist local communities in accessing other sources of finance, for example, grants and soft loans from regional development banks and bilateral and multilateral agencies.

OECD energy-related R&D investments increased after the 1997 oil price shocks, but have declined in real terms since the early 1980s, by as much as 75% in some countries,[42] with the coal and nuclear sectors seeing the sharpest falls. By contrast, corresponding investments in energy efficiency and renewables have seen an increase to around US$2 billion a year, approximately 20% of total energy sector R&D investments. It is important for future development of technology and the transfer of technologies to encourage international collaboration in energy R&D between national governments, the private sector and researchers in academic institutions. In addition to North–South partnerships, South–South partnerships can also be beneficial as locally developed technology is often more readily suited to local conditions. Where developing countries lack investment for technology transfer between themselves, a North–South–South partnership can be beneficial, with industrialized countries providing the investment needed for developing countries to transfer the technology, know-how and skills between themselves. Public–private partnerships are also important for financing renewable energy projects in developing countries. A good example is the development of geothermal power

---

[40] World Bank (1998).
[41] Maurer and Bhandari (2000).
[42] IEA (1997).

Box 3.4: Public–private partnerships for financing renewable energy projects

The public–private partnership has proved for the most part an efficient vehicle for enhancing renewable energy projects in emerging markets. The development of geothermal energy in the Philippines during the 1990s may serve as a spectacular example of this combined approach to successful implementation. In the beginning the Philippines suffered a severe energy crisis, with power shortages and 'brownouts' having a damaging impact on the overall economy of the country. Geothermal power generation was relatively low, some 880 MW, despite the existence of proven potential in the country to use this indigenous resource for enlargement of its generating capacity.

At this point, the GEF agreed to grant US$30 million to enable the Philippine National Oil Company (PNOC) to invest in geothermal field development. The government of the Philippines took it upon itself to implement the necessary legislation. After drilling geothermal production wells in the most prospective geothermal areas, the resource was developed and PNOC, with government guarantees, initiated an international bidding process for the development of geothermal power generating projects under a BOT (build, operate, transfer) scheme. The leading forces in the private sector of the geothermal industry took part in the process: from the United States, firms such as ORMAT, Magma, California Energy and Oxbow; from Japan, Marubeni, Fuji and Toshiba. Private industry played an important role in the process of technology transfer and in the transfer of skills and know-how.

The outcome was impressive. The combined effort of GEF investment and government-initiated legislation triggered investment of about US$1.5 billion to build more than 600 MW of geothermal power plants. The total geothermal power generating capacity of the Philippines has increased since the beginning of the 1990s by more than 1,000 MW, to just over 1,900 MW, and now generates 21.52% of the nation's energy supply. In the process, the power crisis was overcome.

Concerted effort by multilateral agencies and local governments implementing the right legislation can trigger strategic projects, enabling private industry to finance implementation of renewable energy projects in emerging markets.

Source: Lucien Bronicki, ORMAT.

---

in the Philippines (see Box 3.4). If available funds are managed commercially, subject to competitive forces and accountability, projects stand a greater chance of success. Furthermore, combining energy, telecommunications, health, education and economic development in integrated strategies could help to liberate more funding for sustainable energy projects, and the resulting developmental benefits are likely to have a more significant and sustainable impact on the recipient communities than if projects are implemented in isolation.

3.3   Finance mechanisms

3.3.1   *Innovative finance mechanisms*

Existing mechanisms Encouraging investment is only one half of the equation; facilitating access to finance and helping local SMEs to develop the capacity and skills to manage the funds in a sustainable manner are just as important. A number of financing projects and intermediaries have been developed recently which aim to help in this second area by assisting banks and investors in evaluating renewable energy projects, providing business advisory services alongside investment; providing early-stage funding to help SMEs build up a track record (an SME 'incubator'); and channelling private investment to renewables projects. Examples of recent financing projects and intermediaries include the Global Renewable Energy and Energy Efficiency Fund for Emerging Markets (REEF); the Technology Cooperation Agreement Pilot Project (TCAPP); the Solar Development Group (SDG); the African Rural Energy Enterprise Development (AREED) Initiative; the Renewable Energy Technology and Energy Efficiency Investment Advisory Facility (RET/EE IAF); and the Small and Medium Scale Enterprise Programme.

❑ Renewable Energy and Energy Efficiency Fund (REEF):[43]   REEF, which became operational in March 2000, was launched by the World Bank Group's IFC, together with support from the GEF and several other private- and public-sector groups. It is the first global private equity fund devoted exclusively to investments in emerging market renewable energy and energy efficiency projects. REEF had equity funding of US$65 million in its first year, including US$15 million from IFC. REEF actively seeks to make minority equity and quasi-equity investments in profitable, commercially viable private companies and projects in sectors that include on- or off-grid electricity generation primarily fuelled by renewable energy sources, energy efficiency and conservation, and renewable energy/efficiency product manufacturing and financing. REEF's investment criteria are as follows:

---

[43] Adapted from information provided by Dana Younger (IFC) to the G8 Renewable Energy Task Force.

- *technology sectors:* low-impact hydro, wind, solar, biomass, geo-thermal, energy conservation and energy efficiency;
- *geographical focus:* emerging market countries worldwide eligible for IFC financing, including markets in Africa, Mexico and Latin America, the Caribbean, Asia, and central and eastern Europe;
- *investment size:* REEF will consider investment in projects with total capitalization requirements of between US$1 million and US$100 million;
- *instruments:* REEF's investments may take a variety of forms, including common and preferred stock, partnership and limited liability company interests, and convertible or subordinated debt with equity warrants/options. REEF may also make loans to projects or project sponsors on a bridge or permanent basis. Equity transactions are typically structured so that the entrepreneur retains the majority of shares and/or management of the company.

The fund, which is managed by EIF Group, a US-based organization, will be supported by a parallel discretionary debt facility of up to US$100 million consisting of an IFC 'A' loan of US$20 million and up to US$80 million in IFC 'B' loans.[44] The fund will also have access to a unique cofinancing arrangement with up to US$30 million in concessional funds from the GEF. This will allow REEF to invest in smaller and more difficult projects in addition to making its larger commercial investments.

❑ Technology Cooperation Agreement Pilot Project (TCAPP):[45] The Technology Cooperation Agreement Pilot Project was launched in 1997 by three US government agencies: the US Agency for International

---

[44] The IFC offers fixed- and variable-rate loans from its own account to private sector projects in developing countries. These are called 'A' loans. Mobilizing funds from other investors and lenders for private sector projects in developing countries is one of the IFC's core functions: the corporation actively seeks partners for joint ventures and raises additional finance by encouraging other institutions to invest in IFC projects. The cornerstone of the IFC's finance mobilization efforts is the loan participation programme, which arranges syndicated loans from commercial banks, providing additional financing to IFC-financed projects in developing countries. In these syndicated loans, called 'B' loans, participating banks provide their own funds and take their own commercial risk, while IFC remains the lender of record. See <www.ifc.org>.

[45] Information from <www.nrel.gov/tcapp>.

Development (USAID), the US Environmental Protection Agency (USEPA) and the US Department of Energy (USDOE). It is based on longer-term funding and cooperation in the form of programme funding rather than financing shorter-term individual small projects. The aim is to make it quicker and easier to access funds for renewable energy projects and other projects that fit in with the development goals and needs of the host developing country. TCAPP is designed to achieve the following major goals: to foster private investment in clean energy technologies that meet development needs and reduce greenhouse gas emissions; to engage host country and international donor support for actions to build sustainable markets for clean energy technologies; and to establish a model for international technology transfer under the UNFCCC.

TCAPP employs a strategic and collaborative approach to facilitate large-scale international investment in clean energy technologies consistent with the sustainable development needs of developing countries. It is currently facilitating voluntary partnerships between the governments of Brazil, China, Egypt, Kazakhstan, Korea, Mexico and the Philippines, the private sector and the donor community on a common set of actions that will advance implementation of clean energy technologies. This approach is also being used by 14 countries in the SADC with a regional technology cooperation needs assessment that was recently initiated by the Climate Technology Initiative.[46]

The countries participating in TCAPP have made significant progress in developing strategies for building sustainable technology markets and have begun to implement actions aimed at mobilizing private investment and donor support to address country-specific technology cooperation needs.

❑ Solar Development Group (SDG):[47] Drawing on its experience of successful investment in SMEs, where financing is preceded by technical

---

[46] The CTI was launched at the First Conference of the Parties to the UNFCCC in Berlin in 1995, by 23 IEA/OECD countries and the European Commission. Its mission is to promote the objectives of the UNFCCC by fostering international cooperation for accelerated development and diffusion of climate-friendly technologies and practices for all activities and GHGs. The CTI secretariat is based at the IEA in Paris.

[47] Adapted from information provided to the G8 Renewable Energy Task Force by Dana Younger, IFC, and Richard Spencer, World Bank, along with Triodos PV Partners.

assistance, the World Bank and IFC, along with a number of charitable foundations and the GEF, have developed the Solar Development Group (SDG). The SDG is structured to be both a financing window for small PV enterprises in developing countries, which will leverage private sector funds into this emerging sector, and a business advisory service. Its mission is to accelerate the development of viable private-sector business activity in the distribution, retail sales and financing of off-grid rural electrification applications in developing countries, initially concentrating on PV because that appears to be where the greatest demand is at the moment. Formidable barriers, in particular weak physical distribution systems, lack of credit and the high initial cost of off-grid systems, often keep this technically feasible technology beyond the reach even of most middle- and upper-income rural families. The major energy companies that manufacture and develop PV systems have also generally remained marginal actors in the developing world because there are increasingly lucrative developed-world markets which are easier to access.

The SDG has a target capitalization of US$50 million, of which more than US$42 million has been committed. It will consist of two separate programmes: (1) Solar Development Capital (SDC), an investment fund of approximately US$30 million for financing private-sector PV or PV-related companies and financial institutions; and (2) Solar Development Foundation (SDF), which is expected to disburse approximately US$20 million in grants or 'soft' loans both to companies and to programmes that further SDG's mission. A total of 10 local PV companies have already received financial support through SDF and another 12 are expected to be funded during 2001. Over 200 companies in 57 countries have been identified as potential further recipients of support and are under evaluation. Both SDC and SDF are operational and are managed by Triodos PV Partners, a US-based company.

❑ African Rural Energy Enterprise Development (AREED) Initiative:[48] UNEP, in partnership with E&Co (USA), have set up the African Rural Energy Enterprise Development (AREED) Initiative with funding support

---

[48] Information from <www.areed.org>.

from the United Nations Foundation. The AREED initiative seeks to develop sustainable energy enterprises that use clean, efficient and renewable energy technologies to meet the energy requirements of the poor, thereby reducing the environmental and health consequences of existing energy use patterns.

AREED provides enterprise development services to entrepreneurs and early-stage funding, in the form of debt and equity, to help build successful businesses that supply clean energy technologies and services to rural African customers. Such initiatives, which provide early-stage funding to assist SMEs to grow, are sometimes referred to as SME 'incubators'. Services include training, hands-on business development assistance and, for promising businesses, early-stage investment and assistance in securing financing. Providing business development services builds capacity in entrepreneurs, enabling them to reach the level where they can interest a financial institution in considering an investment. Many of the entrepreneurs with whom AREED works would not be able to advance their business ideas without such support because of lack of business knowledge and/or access to early-stage capital. In January 2000 AREED had more than 30 projects in the pipeline.

In each country, AREED works in partnership with a local NGO or development organization to which it will seek to transfer the techniques of energy enterprise development. This is a major project goal, because it is recognized that in order for the AREED enterprise development approach to be sustainable, local capacity must be created to provide long-term support. The main barrier faced in the project to date has been the time commitment required effectively to increase the capacity of local organizations to deliver business development services. Another challenge has been the lack of entrepreneurial experience of many of the African participants. Many of the enterprise activities are at an early development stage, requiring significant amounts of enterprise development services and small amounts of investment capital. Fortunately, the project is structured to be able to provide both.

❏ Renewable Energy Technology and Energy Efficiency Investment Advisory Facility (RET/EE IAF): The facility provides targeted expertise to banks and finance institutions to help them evaluate proposals in the renewable energy or energy efficiency sectors and to develop the skills to evaluate such projects independently. The objectives are to increase investment in commercial RET and EE projects in developing countries, and to build capacity and confidence among financial institutions in the sustainable energy investment sector. The project is run by UNEP with funding from the GEF.

❏ The Small and Medium Scale Enterprise Programme: This programme stimulates greater involvement of private SME in activities eligible for GEF funding by lending GEF grant funds to carefully screened intermediaries over long terms at low interest rates. Intermediaries commit to using the funds to finance GEF-eligible SME projects, through either debt or equity investments. The programme allows the intermediaries to fund long-term loans or equity investments in relatively high-risk, experimental SMEs, for which normally priced capital is presently lacking. By March 2001 US$1.6 million in loans had been approved for three rural PV electrification projects: Soluz Dominicana in the Dominican Republic, Grameen Shakti in Bangladesh and SELCO-Vietnam.

Emerging mechanisms   In addition to these recent financing projects and intermediaries, other mechanisms suitable for promoting investments in renewable energy technology transfer are still in the process of emerging, for example the Clean Development Mechanism (covered in more detail in Section 3.3.2) and the concept of a 'patient capital fund'. Many renewable energy enterprises in developing countries take time to build up their business and often need 'hand-holding' to reach sustainability. Many enterprises have limited business experience and need support not only in preparing a credible business plan but also in project implementation, approaching sources of funding and negotiating agreements with partners. Experience shows that although such businesses provides significant and valuable local services, fundamental to addressing the requirements of the rural population, in many cases they do not meet the litmus tests of mature investment markets. After seed funding has helped SMEs to set up and get started, they find it difficult to access the funds needed to help them expand and establish themselves well enough to qualify for commercial funding.[49] Even when they reach sustainability, not all enterprises will be attractive investment targets for typical investors in the energy market. This is not a poor reflection on the businesses, but rather a market reality. These enterprises deliver energy services on a sustainable basis, but the nature of their business, the rate of growth in demand for their services, the political and cultural setting in which they operate and the products they promote do not necessarily make them acceptable candidates for more commercially-orientated investment. There is a need for 'patient capital' to be invested in these enterprises over the longer term, to help them build up their capacity and track record and become attractive for commercial investment.[50] The 'patient capital fund' should

- offer funds at realistic terms and conditions that will allow the enterprise to meet its medium-term funding needs;
- if necessary accept equity returns of 10% (or less), rather than the more usual target of 20%, though expected returns may reflect market interest rates;
- manage the level of investment risk, which may vary from project to project, through a balanced portfolio approach;
- be managed as a local or regional fund (rather than a window in an orthodox bank) with a management group that understands the needs of the RE market and has the capacity to provide a level of business development support. The fee structure for the management and business development services must be realistic given the low returns on the fund and the fact that many investments will be modest in scale ($250,000 to $1 million);
- recognize that in many cases the enterprises are unable to sustain borrowings and repayments in foreign currency, and seek to offer local currency support wherever possible;

---

[49] E&Co is a US-based non-profit organization that provides business development services and seed capital. It has invested in some 13 PV enterprises globally and each has had to address this hurdle. Some have stalled and are still seeking a solution; others have attracted investment but at significant cost to the company and management, with terms and conditions that could severely restrict the capacity to draw future investment; a few have found a major investor who understands the nature of the market and its long-term potential.
[50] E&Co. is helping to develop the concept of 'patient capital'.

- offer funds that can be accessed with limited traditional security instruments;
- allow the enterprises to enter into subsequent borrowings without unreasonable restrictions – the patient funds would not be considered senior debt;
- establish funds on a long-term basis and look to 'institutionalize' the funds with time; while individual investments may be needed for upwards of five years, it must be recognized that the need within the market for such funds will persist for 10 to 20 years or more.

Conversely, the 'patient capital fund' should not

- assume that the funds are likely to yield short-term, high returns;
- be provided through a local public- or private-sector banking entity that has no experience in the RE field and is, because of standard lending terms, unable to consider offering funds on the required terms and conditions irrespective of the fact that the entity receives concessional investment for this purpose;
- impose terms and conditions on the developer that hamper their ability to raise later-stage funding;
- have agreements and legal conditions that are disproportionately onerous, cumbersome, time-consuming and expensive to execute;
- attempt to apply mature market terms, conditions and contracts to an emerging and immature market;
- set up funds that have a limited life, as the need will be long-term.

It is unlikely that such funds will initially attract private-sector investors or the more commercial interests of the multilateral development banks – though many are keen to provide the later-stage investments when others have brought the risk down. Consideration can be given to the use of guarantees to encourage risk-averse or domestic institutions to provide such investment support.[51]

---

[51] A novel approach along these lines is currently being considered by the IFC, which, with GEF support, will offer financing for the exploratory phase of geothermal projects on a shared risk basis (information provided to the G8 Renewable Energy Task Force by Gunter Schramm of the IFC).

3.3.2 Potential investment via the CDM

The Kyoto Protocol and in particular the Clean Development Mechanism (CDM) could be one way of channelling additional investment towards renewable energy projects. In order to estimate the potential size of investment in renewable energy technology transfer to developing countries from the CDM, several calculations need to be made: an estimate of the size of required emissions reductions; an estimate of the size of the CDM; an assessment of the role of renewables in the Kyoto mechanisms; and an estimate of the likely prominence of renewable energy projects in the CDM.

Estimated size of required emissions reductions   The Kyoto Protocol commits a group of countries (listed in Annex B of the Protocol) to reduce annual emissions of greenhouse gases by 5.2% overall in the period 2008–12, compared to 1990 levels. Analysis by the IEA suggests that, allowing for the expected business-as-usual growth in emissions from the OECD countries and the recent reductions in energy use of the EIT, there will need to be a reduction in the projected levels of annual emissions in 2010 of over 2,100 Mt carbon dioxide to meet the Kyoto targets.[52] This is equivalent to an overall 16% reduction in carbon dioxide emissions (see Table 3.4). If the required reduction of 2,100 Mt carbon dioxide were to be met entirely through the deployment of renewables for power generation, then this would equate to something in the region of 930 GW of additional renewable energy generating capacity over the period 2000–10.[53] This is about 70% of the total new electricity generating capacity that *WEO* 2000 has estimated will be required globally over the period 1997–2010 (1,335 GW),[54] and dwarfs the likely estimated increase in non-hydro new renewable generating capacity over this period (43 GW).[55] It is also

---

[52] *WEO* (2000).

[53] Based on calculations carried out by AEA Technology Environment (Simmonds and Taylor 2000), but using updated *WEO* 2000 figures. Figures assume that the additional capacity of new renewables is built up in linear fashion over the period 2000–10 and displaces CO_2 emissions from the projected global average generating mix in 2010. The renewable energy systems are assumed to be of a global average mix with average load factors.

[54] See Table 1.4.

[55] See Table 1.3.

Table 3.4: Carbon dioxide emissions and targets in Annex B countries, 2010 (Mt CO_2)

Country	2010 target emissions	WEO projected 2010 emissions	Gap (%)[a]
OECD North America	4,935	6,995	41.7
OECD Europe[b]	3,664	4,323	18.0
OECD Pacific	1,307	1,682	28.7
Russia	2,357	1,670	−29.1
Ukraine and eastern Europe	1,150	867	−24.6
Total	*13,413*	*15,537*	*15.8*

[a] The gap has been calculated by expressing the difference between target emissions and projected emissions as a percentage of the target emissions. Thus it indicates the extent to which the projected emissions exceed the targets.
[b] Turkey is not included.
Source: WEO (2000).

considerably larger than the total amount of additional capacity needed by OECD countries before the year 2010 (488 GW).[56]

Clearly, an additional 930 GW of renewable energy generating capacity over the period 2000–10 is much larger than would happen in practice, but there are some important observations that can be drawn from this analysis. The GHG emissions commitments agreed at Kyoto are sufficiently ambitious that they have the potential significantly to increase the need for more energy-efficient power-generating technology and fuel switching (including renewable energy generating capacity). For new renewables to have a significant role in helping to meet reduction commitments, there will need to be development of substantial renewable energy electricity generating capacity outside of the OECD countries – for example, in developing countries in Asia that require an increase in generating capacity. The CDM could play a major role in achieving this.

Therefore, realizing the full potential of renewable energy markets under the Kyoto Protocol will depend on the importance accorded to the mechanisms for implementing the Kyoto Protocol (particularly the CDM, as it allows for projects in developing countries where generation capacity is needed), and the attractiveness of renewable energy,

---

[56] See Table 1.4.

compared to other technologies, as an option within each of the Kyoto mechanisms.

Estimated size of the Clean Development Mechanism   A number of studies have estimated the potential size of the Clean Development Mechanism. Most of these studies used purely economic analysis, basing their conclusions on models that project the need of industrialized countries to reduce GHG emissions below business-as-usual levels over the period 2008–12. They then calculate the costs of achieving these reductions through domestic action as compared with using the Kyoto mechanisms. With the details of the CDM not yet agreed, it is difficult to be certain just how large a mechanism the CDM will be and, equally important, how much projects will contribute to the sustainable development goals of host countries. This uncertainty is reflected in the range of values presented in Table 3.5. Estimates of the potential size of the CDM range from 99 to 2,650 Mt carbon dioxide, equivalent to 5–125% of the estimated reductions that industrialized countries will need to achieve.

In reality, the CDM is a project-based mechanism and it is not certain that sufficient projects could be implemented to generate thousands of millions of tonnes of emissions reductions in the short time available before the end of the first commitment period.

Work by Vrolijk using the RIIA's emissions trading model suggests that the size of the CDM is likely to fall in the range of 246–517 Mt carbon dioxide (8–16% of reductions needed to meet commitments).[57] The potential volume and value of CERs from CDM projects is important to Annex I parties investing in the projects. However, of more importance to the recipient developing countries are the magnitude and direction of CDM investment flows to them. This again is subject to considerable uncertainty, but Grubb and Vrolijk estimate the potential CDM market size in the range of US$5–10 billion per annum investment.[58] A recent study by AEA Technology estimated that the

---

[57] Vrolijk (1999). This low estimate for the CDM is based on a large estimate of 'hot air' and 'no regrets' measures (measures that limit emissions of GHGs, below levels that would otherwise be achieved, at no net cost).

[58] Grubb and Vrolijk (1998).

Table 3.5: Estimates of the size of the CDM

Study	Emission credits (MtCO₂)
Haites	99–2,097
MIT	1,001–2651
Austin	1,456–1,844
US administration	367–689
RIIA	246–517

Source: Vrolijk (1999).

RIIA model range for the likely size of the CDM would translate into additional capacity of renewable energy generating technologies of between 105 and 220 GW, or around US$130–275 billion of capital investment over the period to 2010.[59] For the purposes of this book, investment of US$275 billion (or US$28 billion per year) will be taken as the likely upper limit to the impact that the CDM will have on promoting additional renewables capacity.

A report by the Ad Hoc Working Group on the CDM (UN, 1999) estimates the size of ODA flows at around US$50 billion per year and FDI at around US$240 billion per year.[60] Thus the CDM could generate significant additional investment by industrialized countries in the developing world by increasing current investment by around 10%. However, there is some debate as to whether, in some cases, it may displace existing FDI, resulting in no change in the overall level of flows but perhaps changing the type of activity that is being funded to one more closely targeted at achieving sustainable development.

The role of renewable energy in the Kyoto mechanisms There is a range of options with which countries can fulfil their Annex B commitments. The Kyoto Protocol makes clear that the focus of attention should

---

[59] Simmonds and Taylor (2000). It is assumed that the additional capacity of new renewables is built up in linear fashion over the period 2000–10 and displaces CO₂ emissions from the projected global average generating mix in 2010. The renewable energy systems are assumed to be of a global average mix with average load factors. An average investment cost per GW of new renewable energy capacity is calculated on the same basis.
[60] UN (1999).

be on domestic actions, which include the following: encouraging energy efficiency; reducing the demand for energy services; reducing non-carbon dioxide GHGs; and encouraging fuel switching to less carbon-intensive fuels, including the use of renewable energy sources. The Kyoto mechanisms also provide the opportunity for reducing emissions by participating in emissions trading or investment in projects in other countries that achieve reductions in GHGs. Assuming that the structure and operation of the CDM can be successfully set up and outstanding issues resolved, this is likely to offer considerable potential for enhancing the renewable energy market, because developing countries have a strong need for additional power generation capacity. Also, the requirement that projects assist in the sustainable development of these countries is likely to favour renewables over conventional sources of power generation.

The attractiveness of renewable energy under the Clean Development Mechanism   Renewable energy technologies will have to compete with a wide range of other options for funding under the Clean Development Mechanism. Only nuclear projects have been ruled out of the CDM, so it is likely that renewables will need to compete with options such as energy efficiency; fuel switching; cleaner fossil generation technologies; methane gas capture; and afforestation and reforestation projects. Clearly, potential investors will be looking for projects that are the most cost-effective in reducing emissions of GHGs, while host countries will want to ensure that projects meet their sustainable development objectives. Projects that fulfil both these requirements will therefore be most likely to gain CDM funding.

Although new renewable generating technologies may not be the cheapest abatement options in developing countries in certain circumstances, they have three potentially important advantages over other options when being considered for CDM funding:

- Non-Annex-I countries require new generating capacity in order to meet their development goals. Cheaper options such as gas capture, afforestation, reforestation and even energy efficiency will not provide the required increase in capacity.

- New renewable technologies can contribute to a sustainable development path in a way that other electricity generation technologies based on fossil fuels cannot.
- New renewable technologies are well suited to meet energy service requirements in remote rural areas. It has been agreed that small-scale renewable energy projects, which are suitable for remote areas, may be fast-tracked under the CDM.

As long as new renewables can generate electricity at a similar cost to fossil-based alternatives, they may not need to be necessarily the cheapest abatement option in order to take a significant share of CDM funding, particularly if some value is given to the development benefits of renewables. One of the major non-cost benefits that renewables have over other forms of electricity generation and other possible CDM projects is their contribution to sustainable development. There are any number of possible definitions for what constitutes sustainable development and as many criteria by which to measure it. However, in the context of the CDM, certain factors such as environmental benefits (other than GHG reduction) and development additionality are often cited as being of particular relevance.

It is not certain how tightly the rules governing the sustainability criteria will be drawn, and hence whether projects selected under the CDM will unambiguously be considered as contributing to 'real' sustainable development. Technologies such as large hydro and clean coal technology may qualify for funding under the CDM as they reduce GHG emissions, but some NGOs are questioning their benefits regarding sustainability. Even among less controversial technologies, such as the use of renewables for rural electrification or the promotion of more energy-efficient compact fluorescent light bulbs, there is some debate as to which are likely to bring the stronger sustainability benefits.

The development benefits of CDM projects will depend on the country context, project type and how each project is implemented in terms of local participation, capacity building and technology transfer. The value placed on these benefits will vary depending on their relevance to achieving development goals stated in the NSSDs and PRSPs of individual countries. Sustainability indicators are needed to help assess the relative developmental benefits of projects and contributions towards different aspects of sustainability such as prospects for technology transfer; contributions towards poverty alleviation; potential for local capacity building; and potential local and global environmental effects.[61] Ultimately it will be the host countries that will decide if projects contribute towards their sustainable development goals.

The likely proportion of the CDM taken up by renewable energy projects The AIJ programme is the closest mechanism to the CDM currently in existence. One way of estimating the likely proportion of renewable energy schemes in projects under the CDM is to look at the trend in selection of projects for AIJ investment, some of which are in developing countries (CDM type projects) and some of which are in EIT (JI type projects).

Figure 3.1 shows that of the official 140 AIJ projects reported to the UNFCCC by June 2000, 49 projects or 35% involve renewable energy. This compares to 50 energy efficiency projects, 13 forest preservation, reforestation and restoration projects, and 6 projects involving fuel switching.

However, Figure 3.2 shows that because the savings from individual renewable energy projects tend to be small, the total contribution of renewable energy to emissions reduction under AIJ is only 17%. If this share of the savings were achieved by renewables under CDM, then, using the above figures for the total CDM market (105–220 GW), this would equate to an additional 18–37 GW of renewables capacity.[62] This is equivalent to a capital investment of US$23–47 billion over the period 2000–10 (or around US$2.3–4.7 billion per year).[63] Table 1.3

---

[61] Begg et al. (2000).

[62] Assuming that the additional capacity of new renewables is built up in linear fashion over the period 2000–10 and displaces CO_2 emissions from the projected global average generating mix in 2010. The renewable energy systems are assumed to be of a global average mix with average load factors.

[63] Assuming the renewable energy systems are of a global average mix with average load factors. An average investment cost per GW of new renewable energy capacity is calculated on that basis.

Figure 3.1: Proportion of project types under AIJ (%)

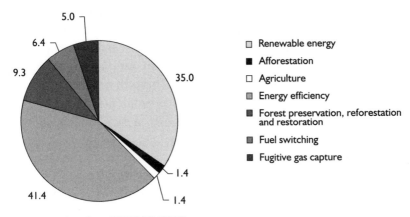

Source: Data taken from UNFCCC (2000).

Figure 3.2: Contribution to emissions savings under AIJ (%)

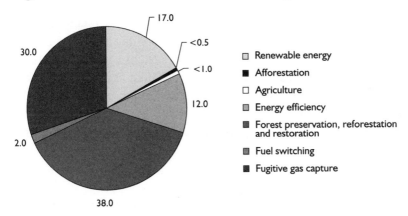

Source: Data taken from UNFCCC (2000).

shows that non-hydro renewables are predicted to increase by around 43 GW from 1997 to 2010; the additional increase from the CDM could bring the total increase by 2010 up to around 77 GW, almost doubling the projected increase from non-hydro renewable energy generating capacity. Hence the CDM has the potential to bring very significant investment to new renewable energy projects in developing countries.

The most economically attractive projects under the CDM will not necessarily be those with the strongest sustainability benefits. Therefore, the methodology and criteria by which projects are actually selected and agreed, and the relative weight attached to economic and development criteria, will have a significant impact on the type of projects which are funded under the CDM. This will have a big influence on where renewables will sit within the estimated range of an additional 18–37 GW likely for new renewable energy capacity between 2000 and 2010.

Potential benefits of CERs In addition to increasing the overall level of investment for renewable energy systems in developing countries, there is also the potential for CERs to reduce the overall costs of renewable energy systems under the CDM. Accurate estimates of the actual impact cannot be made until the market price for trading of carbon is established. Current predictions range from as little as US$4 to as much as US$100 per tonne of carbon dioxide. A recent study estimated that for typical 40 Wp solar home systems and a market price of US$20 per tonne of carbon dioxide, each system would generate CERs of US$6 per year, which over a 20-year lifetime at a 10% discount rate would equate to about 10% of the system's initial wholesale equipment cost.[64] Alternatively, it has been estimated that the CER of US$6 per year could augment annual fee-for-service revenues by about 3–5%, potentially enabling an otherwise marginal business to attract sufficient investments to develop and grow to a sustainable size.

Conclusions Although it may be that none of the estimates for the size of the CDM, the potential for renewables within the CDM and the impact of CERs on renewable energy costs is accurate, it is generally agreed that the CDM has the potential to channel billions of dollars from industrialized to developing countries. As such, it could fundamentally affect the development of the latter's energy systems, offering them access to the latest 'clean' technologies and thus promoting a more environmentally sustainable path for their economies.

---

[64] Kaufman (2000).

However, before such a vision can be realized, agreement has to be reached on many issues regarding the operational arrangements of the CDM. Whether and how these issues are resolved will determine the attractiveness of the CDM to both potential host countries and donor organizations. If the rules and modalities are not carefully thought out, with climate and other environmental and social impacts considered, then the CDM will fail to meet its twin objectives of helping developing countries to achieve sustainable development and assisting industrialized countries in meeting their emission commitments. Local capacity will need to be built in the host countries to assess the costs and benefits relevant to their respective situations and to select projects for implementation.

For the project selection criteria of the CDM to be most favourable for renewable energy technology, there needs to be an emphasis on sustainable development and meeting the development goals of the recipient developing country when projects are chosen. The development goals should be in line with NDDSs and PRSPs; use of sustainable development indicators could help in selecting projects. There also needs to be an emphasis on indigenous sustainable energy resources to provide a more secure energy mix for the future (to prevent fossil-fuel-based energy efficient projects from dominating CDM investment). However, there will need to be some compromise on the projects which are allowed, and the views of both the investors and recipient countries must be taken into consideration when coming to an agreement. The balance agreed between the economic and sustainability criteria used for project selection could hold the key to determining the impact on investment for renewable energy projects under the CDM in the future.

Chapter 4

Barriers and Options

4.1 Barriers

4.1.1 Introduction

If renewable energy technologies are transferred successfully and renewable energy systems integrated sustainably into developing countries, benefits will accrue both for the developing country, in respect of social, economic, environmental and sustainable development, and for the world environment.

There have been many renewable energy demonstration projects in developing countries. Some have been successful and others have failed. There is a multitude of factors that influence the success or failure of renewable energy systems.

The term 'barrier' is often used to refer to factors that impede the adoption of a new technology. There is much discussion over the precise definition of the term 'barrier', and in some cases people prefer to refer only to market failures. The term is used fairly broadly in this book to refer to any technical, economic, institutional, legal, political, social or environmental factor impeding the deployment of renewable energy technologies. Barriers tend to be interrelated, as shown in Figure 4.1, so it is often difficult to isolate the impact of any one barrier in particular.

There is a good resource base and potential for the development of renewable energy systems in developing countries. However, barriers are preventing the full potential of renewable energy technology from being reached, and slowing down the process of technology transfer. In order to realize the full potential of renewable energy systems it is necessary to identify and address the barriers to technology transfer which are preventing the successful implementation of renewable energy systems. In many cases the barriers to technology transfer are similar in industrialized and developing countries, although specific national

Figure 4.1: Interrelationship of barriers to renewable energy technology in developing countries

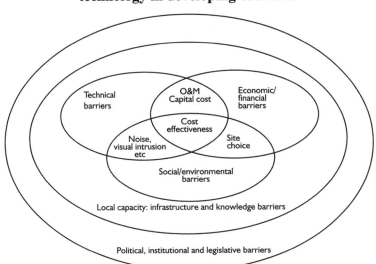

Source: Adapted from Smith and Marsh (1997). 'Local capacity: infrastructure and knowledge' has been added as this is a significant problem in developing countries in contrast to industrialized countries.

characteristics play an important role in determining the barriers within each country, and local characteristics impact on the barriers for particular projects. Some of the barriers in developing countries are symptomatic of lack of development, while others are more symptomatic of weak policies and lack of government targets.

The barriers in developing countries attributable to their state of development include:

- investment risk (real and perceived) due to political and economic instability and corruption;
- lack of government agency capability to regulate or promote technologies;
- lack of energy and land use plans and of planning capability in general;
- lack of management experience;
- low technical capabilities of domestic firms;
- risks to recipient country (such as loss of indigenous industry or cultural drawbacks).

There are also barriers when transferring technology from one country to another, especially when there are cultural differences and different energy service requirements; these include:

- problems of bridging cultural and language gaps and fostering long-term relationships;
- missing connections between potential partners;
- lack of understanding on the part of suppliers of the requirements or priorities of technology recipients;
- lack of understanding of domestic conditions by foreign firms;
- insufficient investment in R&D, particularly for adapting technology to local conditions.

One of the fundamental barriers which is often faced in transferring technology to developing countries is that the technology being transferred is not appropriate to the local context and demands or is not adapted to the local environment. It is important to understand the local situation in developing countries and meet the required energy service demands with appropriate technology. The right combination of energy sources and technologies must be identified for each situation. This might mean a combination of energy-efficient technologies powered by a range of renewables and fossil fuels, providing a range of energy services.

The main barriers faced in renewable energy technology transfer to developing countries are outlined under the headings set out in Table 4.1. By no means every barrier will be faced by each renewable

Table 4.1: Barriers facing renewable energy technology

Type of barrier	Headings
Political, institutional and legislative	National policies and programmes
	Institutional structures
	Intellectual property and standards
Local capacity: infrastructure and knowledge	Information exchange, education and training
Economic/financial	Financing
Social/environmental	Social
Technical	Other

energy project. Only a few barriers may be relevant for the particular situation faced; but in order to identify the key barriers early on and address them before they become a problem, it is important to be aware of the range of possible barriers. The categories below can be used as a kind of checklist against which to identify barriers for a particular project. It is important to understand which barriers may pose a significant threat to the success of a project if ignored and which are easily dealt with. Learning from past experience the best way to overcome certain barriers is crucial to help replicate best practice. This topic is dealt with in more detail in Section 4.3.

4.1.2 National policies and programmes

The main barriers to renewable energy technology transfer that are related to national policies and programmes are:

- lack of clear government plans and targets for renewable energy development;
- lack of appropriate fiscal policies and support mechanisms (taxes, duty, pricing, etc.);
- unclear and changing grid electrification plans;
- lack of access to the grid;
- lack of integrated planning for energy and development;
- lack of consistent policy;
- lack of focus and ownership for renewable energy development.

Government plans and targets   If the government has clear plans or targets for renewable energy implementation, then this is a signal to other key players that the government is committed to renewable energy development. Without this government commitment, investors, suppliers and developers are apprehensive that adequate support may not be made available (in the form of policies, incentives and supporting infrastructure) and are less likely to invest time and money in developing a market in that country. Government plans and targets are given more weight if favourable policies and government-funded renewable energy programmes are also put in place.

Fiscal policy and support mechanisms   In many cases there is inappropriate taxation and duty on imported renewable energy technology components and direct-current appliances. For aid-funded projects, renewable energy technology is imported free of duty and tax, but in some cases duties are as high as 30–40% after the initial installation, so obtaining spare parts becomes too expensive, unless locally manufactured parts can be used.

All over the world rural energy systems are subsidized (in both industrialized and developing countries). Governments will usually subsidize something only if there is a social or environmental benefit to be gained thereby. For example, it is common in industrialized countries to charge all households the same fee for electricity whether they are in an urban or rural area. In effect, this is a cross-subsidy from urban areas to reduce costs in rural areas where it is more expensive to provide power. In general, energy subsidies have been shown to benefit the rich more than the poor, as the rich tend to use more energy. It has therefore been recognized that subsidies in general are not necessarily a good thing, unless the subsidy can target the poorest people ('smart subsidies') and is sustainable or can be removed without destroying the market.

Grid electrification plans   In order for developers to estimate the potential market for stand-alone renewable energy systems, or mini-grids, it is important to know which areas are to receive grid electrification within the next 5–10 years. With this knowledge they can plan effectively and economically which areas they should consider for mini-grids and which areas might benefit from pre-electrification systems. Pre-electrification systems are designed to give access to electric power for a few years only while the area is waiting for grid connection. Once the grid arrives, the systems can then be moved on to another village.

Access to the grid   The installation of stand-alone industrial renewable energy power systems is strongly influenced by the opportunity to sell excess power to the grid and therefore the 'access' such systems have to the grid. The introduction of small power purchase agreements (PPAs), which formulate the rules and regulations for small power-producing industries to sell power to the grid, is a step in the right direction. The details of the agreement regarding the purchase tariff structure and the amount of electricity the grid will buy is crucial. In some cases the agreements are on a 'must buy' basis, i.e. the grid guarantees to buy all the excess power the industry produces. There are also variations in tariff for firm power (guaranteed production) and non-firm power (variable levels of production). The economics of installing stand-alone renewable energy systems in rural industries (e.g. palm oil mills or sugar mills) depends heavily on the price and conditions of exporting electricity to the grid, or the existence of a local community which will pay for electricity from a local mini-grid.

Lack of integrated planning Developing countries would do well to integrate energy issues into their development planning. For example, when schools or health clinics are built, the energy requirements of these facilities should be planned on a sustainable basis, rather than bolted on afterwards.

Lack of consistent policy Many technologies may depend on the development of policies and regimes that provide an appropriate framework. If there is not a consistent policy for renewable energy, then this can increase political risks and uncertainty and also raise the transaction costs.

Lack of focus and ownership for renewable energy development It is often the case that several different and poorly coordinated departments are involved in renewable energy projects. It is important instead to have one government department or organization responsible for the development of renewable energy systems in each country: a 'champion' for renewable energy.

4.1.3 Institutional structures

The implementation of renewable energy systems is affected by policy and regulations relating to rural development, rural electrification, planning and environmental protection. In most cases the responsibility for these areas is divided among several government departments and the electricity utility. Communication and coordination among those responsible for these areas are often poor, which is not conducive to a smooth path for technology transfer. If the responsibility for renewable energy policy and planning is split between departments, this can result in a lack of commitment to push forward the revised policy and planning regulations needed to aid technology transfer. The process of getting planning approval for renewable energy systems can be a long-drawn-out process, so that by the time the approval is given (which can be up to a few years later), the energy requirements or circumstances may have changed.

4.1.4 Intellectual property and standards

The main barriers to renewable energy technology transfer that are related to intellectual property and standards issues are:

- weak or unclear law on intellectual property rights;
- lack of supporting legal institutions;
- lack of technical standards and quality control.

Technology imported from industrialized countries is in general more expensive than technology manufactured locally in developing countries, mainly due to the higher labour costs and higher quality standards prevailing in the former. Many developing countries are therefore keen to set up a local manufacturing base via a joint venture. There are two main barriers that prevent this from happening: the fear on the part of manufacturers that their technology designs may be stolen or otherwise compromised, thus preventing them from recovering their R&D investment costs, and the lack of sufficiently high local technical standards, resulting in poorer quality of production.

Intellectual property rights (IPR)   Most buyers of technology view it as a functional item they gain, but for most manufacturers technology is the design and engineering behind the product. This is an important distinction of perspective when trying to understand the issues involved with IPR.

The term intellectual property (IP) is relatively recent. It was created as a shorthand phrase encompassing both the traditional concepts of industrial rights (patents, trade secrets and trademarks) and artistic rights (copyright). IPR laws are different from country to country, but they have the following general features: they define the standards for what is and is not protectable; they provide procedures that must be followed to obtain protection; they define the scope and duration of protection; and they establish the consequences for anyone who infringes the protected right. There is some international agreement and harmonization among countries, but IPR remain under national law. The World Intellectual Property Organization (WIPO) seeks to foster and harmonize IPR protection among industrialized and developing countries and to disseminate information. The 1994 Agreement on Trade Related Aspects of Intellectual Property (TRIP), negotiated in the context of the Uruguay Round of the General Agreement on Tariffs and Trade (GATT), is leading to increased homogeneity of laws around the world in accordance with minimum standards.[1]

IPR provide both advantages and disadvantages for renewable energy technology transfer. The advantages which might result from robust IPR include an increase in innovation due to the rewards IPR provide; fair treatment of innovators; public disclosure of patent technologies (at a cost); sharing of secrets under confidentiality agreements (at a cost); ease of purchase, sale or licence; and enhanced investment as a result of the assurance that investors will reap rewards.[2] But robust IPR alone are not enough to guarantee investment; other conditions need to be in place too, as mentioned under other headings in this chapter. Indeed, if IPR are too strong, they can create barriers to technology transfer, including restricted access to technology; increased cost to users; monopolization and centralization of commercial and technical advantage in the hands of those who conduct research, can afford to protect IP and are knowledgeable in how to deal with it; restrictions on sharing information and scientific cooperation; and the national institutional burden of processing IPR. There is a risk that

---

[1] IPCC (2000), ch. 3.
[2] Ibid.

excessively strict IPR adversely affect follow-on innovations and slow down the pace of technology development.

Barriers can also be created by poor IPR as companies hide important information for fear of misuse. This can drastically impede the adaptation of renewable energy technology to local conditions. Also, unprotected technology is not attractive to investors, and this will reduce the chances of technology being transferred, developed and manufactured locally. It is important to get the right balance with IPR, such that they are clear, and will encourage investment, innovation, partnerships and the transfer of technology. In many developing countries, the law regarding IPR is not very specific, nor does it offer manufacturers much protection. The enforcement of such law may also be rather weak. These shortcomings deter many companies from setting up factories in developing countries to manufacture their systems; alternatively, factories are set up that manufacture older technical designs on which patents have expired or are near to expiry. If developing countries are to benefit from using the latest technical products, the manufacturers must be allowed to make a return that adequately covers the costs of R&D.[3]

Lack of supporting legal institutions   Weak legal institutions in a host country can be a serious barrier to technology transfer. There are three broad types of risk that actors are exposed to if legal institutions are weak. These are:[4]

- *Contract risk*, covering the likelihood and cost of enforcing contractual legal obligations with other actors. For example, investors will be deterred if weak legal institutions in a host country may mean that they cannot enforce contracts or recover costs through the courts if problems occur.
- *Property risk*, covering asset ownership and corporate governance. For example, if manufacturers can not protect their intellectual property rights, they will either avoid transferring technology or

---

[3] Goldemberg (1992).
[4] IPCC (2000), ch. 4.

transfer only older technologies that put less of their capital stock at risk.

- *Regulatory risk*, covering licensing, tariff setting, taxation, foreign exchange and trade controls. For example, if PPAs or electricity export tariff rates can not be clearly and transparently set and enforced or guaranteed, then actors will be deterred from investing in and building power plants that export electricity to the grid.

Also, legal institutions that mandate financial disclosure by potential partners may not exist or be enforced, so that it is more difficult for potential partners to assess each other's financial condition, increasing potential risks when entering into partnerships and investing.

Standards and quality   Absence of technical standards and quality control is a barrier to local manufacture. Poor technical standards and resulting higher failure rates can damage the reputation of equipment manufacturers and suppliers and their technology. If the quality control is lower, then manufacturers are unable to provide the same warranties on equipment, again damaging their own reputation and that of the technology. Confidence in the technology is important in order to establish a market. If confidence is destroyed, then financiers, dealers, installers and end users will be unwilling to invest in the technology.

Quality and standards are also important to aid technology choice. End users, installers and developers need to be able to make informed choices on a quality versus cost basis. Quality standards, enforced effectively, can help to guard against very poor or even non-functional or fake technology. However, if standards are set too strictly, it can inhibit or destroy market development.

4.1.5   Information exchange, education and training

The main barriers to renewable energy technology transfer that are related to information exchange, education and training issues are:

- lack of access to information;

- lack of skilled local labour and capabilities;
- lack of exchange in ideas and experiences.

Access to information   In order for technology transfer to happen successfully, there first needs to be a clear idea of the potential market for the use of that technology and a desire for the services it provides. There is often a lack of accurate information on the potential renewable energy resources available, and the energy requirements of communities and businesses are poorly understood by foreign organizations. Also, if potential end users are not aware of the services that the technology can provide, there will be no demand for them.

Potential energy system installers and developers are often not familiar with renewable energy technology and do not know the best applications or constraints of different technologies, so would not be able to select the most appropriate options for particular service requirements and local conditions. When technology is selected inappropriately, it is not likely to provide satisfactory energy services and may fail, disappointing users and giving the technology in general a bad image. Potential dealers, manufacturers and installers are often unaware of one another, and so do not make the links necessary to establish partnerships and engage collaboratively in the technology transfer process.

Lack of information regarding quality and standards of technology can be a problem as it means that users, installers and developers are not able to distinguish between good and bad equipment and to make informed choices.

Sometimes there is a lack of confidence in the credibility or objectivity of available information in developing countries regarding technologies, their appropriate applications and the costs associated with them. There is also a lack of confidence in information provided to potential actors in the technology transfer process on market potential in developing countries, willingness and ability to pay, and the energy service requirements of potential users.

Skilled labour   Many technical failures are a result of an absence of indigenous capability. For example, inappropriate choice of technology is often the result of poor searching, selection and negotiation.

Joint ventures may not produce long-term learning, absorption of technology or development of competence if capabilities for learning and integrating are absent or inadequate. Capacity building is needed to assess, select, import, develop and adapt appropriate energy technologies in developing countries. A basic level of indigenous technological capability is essential to facilitate the process of sustainable technology transfer to developing countries' markets.

Lack of local technically trained staff to install, operate and maintain renewable energy equipment is often a problem, particularly in remote rural areas that are difficult to access and where transport is infrequent. If the equipment does not receive the required maintenance regularly it is likely to fail. Often, locally trained staff migrate to urban areas to exploit their new-found skills, so selection of trainees who are likely to remain in the area is an important consideration. Lack of business skills in SMEs can reduce their ability to prepare business plans and market their products and services.

Exchange of ideas and experiences Lessons learned from pilot and demonstration projects are often not analysed and used to steer best practice, or disseminated to other relevant actors (e.g. potential developers, installers, manufacturers, NGOs, users and financiers). Renewable energy technologies from foreign countries are not always directly suitable for the local conditions found in developing countries (e.g. the type of biomass may be different; the wind regime may be more severe, perhaps including typhoons and sandstorms; the annual fluctuations in river level and silt content may be too extreme; etc.). Research and development to adapt the technologies to suit local conditions is often lacking.

4.1.6 Financing

The main barriers to renewable energy technology transfer that are related to financing issues are:

- lack of access to capital;
- lack of investment;

- inappropriate subsidies;
- scale of systems;
- size of organizations;
- dispersed nature of the projects.

Access to capital   In order for installers and end users to purchase, install and use renewable energy systems, they need access to finance. For installers, the main barriers are the regulations and conditions imposed by lenders before approving a loan. For example, many of the installers will be new companies and will have limited records of their accounts and assets; yet lenders may require substantial evidence of track record showing the company to be creditworthy.

For potential end users, the financial barriers are different. Renewable energy systems are in relative terms very expensive for them. In many areas suitable micro-financing packages are not readily available, and even if they are, end users may find it difficult to make regular monthly payments, as their income will in many cases be irregular, depending for example on the harvest of crops or the sale of handicrafts. In remote areas trade may not be carried out regularly each month due to the distances to markets.

Investment   Investment is crucial if renewable energy systems are to become established. Lack of investment can be due to a lack of understanding of the investment profiles and life-cycle costs for renewable energy systems (i.e. a higher up-front cost with longer-term benefits). It can also be due to the use of inappropriate project appraisal methodology to assess the costs and benefits of renewable energy projects compared to those of conventional energy projects. Also, subsidies on other forms of energy technology or fuel may make it difficult for renewable energy systems to compete on an equal basis, rendering the technology unattractive to investors.

Real or perceived risk is a major deterrent for investment. There are many factors that can contribute towards the real or perceived risk of investment in renewable energy systems, including imperfect markets in host countries; uncertainty regarding the political and economic stability of the host country; new technology that is not yet proven on a commercial scale or in host country conditions; and uncertainty over energy demand and market development.

Some estimates predict that in the order of 50,000 rural SMEs will be required to provide access to clean, reliable sources of energy for the majority of the 2 billion people who currently lack it.[5] The number of such enterprises today is in the hundreds rather than the thousands, so there is a large task to be done in expanding existing SMEs and encouraging the establishment of new ones. Investment is therefore crucial, both in existing SMEs to help them expand and in new ones to provide start-up finance. It can be difficult for SMEs to attract investment when they are not yet large enough to show profitability; here SME 'incubators', programmes set up to provide early-stage finance for SMEs to expand after their initial establishment, have an important role to play.

Subsidies   'Hidden' subsidies for fossil fuels in host countries can be a barrier to renewable energy systems, making it difficult for them to compete economically. These may range from subsidized kerosene and coal prices for consumers to government investments in costly grid extensions that are not fully recovered from consumer electricity rates. Subsidies are often implemented where governments see a social or developmental need. However, energy subsidies are often targeted ineffectively and disproportionately benefit rich people, who tend to consume more than the poor whom the subsidies are intended to assist.

In some cases subsidies are applied to renewable energy systems for a fixed period of time. When they are removed, the users, who have been receiving the energy free or at very low cost and think the government should continue to subsidize the energy, often are unwilling or unable to pay; consequently the projects fail or at the very best fail to expand, deterring future investments. In addition, subsidized projects, often funded by multilateral or donor aid, distort local commercial markets and can harm or destroy local industry. If communities get used to receiving aid-funded projects, they may wait for the promise of a subsidized project and delay purchasing technology. 'Smart' subsidies can be developed that are targeted, time-limited and transparent, and benefit the poor.

---

[5] E&Co. (2000); Allen (2000); Gunaratne (1999).

Scale of systems   Many renewable energy technologies operate on a smaller scale than more conventional fossil-fuel technologies. For example, few renewable energy projects exceed US$50 million, whereas most conventional power technologies are substantially larger than this. There are certain fixed costs that are relatively independent of project size, so a reduction in project size increases the transaction costs as a percentage of the overall project cost. This makes it much harder to attract private investment.

Multilateral and bilateral development assistance has traditionally been for large-scale projects, and the agencies that dispense such assistance are not geared up for small financing packages, micro-financing or bundling of projects.

Size of organizations   Relatively large investments are typically needed to manufacture and export new technologies. Renewable energy technology companies are small compared to oil and gas technology companies and in many cases lack the financial strength required to invest in large-scale manufacturing and exports.

Dispersed nature of the projects   The geographically dispersed nature of many renewable energy projects creates inherent difficulties in installation, maintenance and collection of fees in rural areas, because of the distances that have to be travelled and the difficulties with and cost of transport and communications. Installations for a particular area will need to reach a critical mass before the costs of a local engineer and/or fee collector can be covered.

4.1.7   Social

Social acceptance of renewable energy technology is very important, as its absence can be a major barrier. If the local community does not accept the technology there will be no demand for its services. For example, if it is culturally or socially unacceptable in a community for women to cook in the middle of the day instead of in the evenings (for whatever reasons), then solar cookers will not easily be accepted into that community. Women gather and use most of the energy used in households in rural communities, so introducing new energy technologies will impact most on their daily routine and practices. The gender aspects of renewable energy technology need to be understood by those supplying the technology.

Community and end-user involvement in planning projects and selecting technology is often lacking. Such involvement is crucial to help gain social acceptance of the projects and systems and to ensure the technology is providing the required services and addressing the community's priorities. The hierarchy and decision-making process in the community needs to be understood and respected to ensure that community leaders are allowed the opportunity to approve projects and activities. The social fabric of the community needs to be understood when selecting technicians and fee collectors; these people, organizations or groups must be respected by the community if the projects are not to face difficulties.

The local culture, religion and superstitions need to be understood when projects are planned, in order to avoid problems later in the development stages. The ability and willingness of the community to accept practices and concepts that may be alien to their culture, such as dealing with cash or taking out credit, also need to be assessed.

A lack of entrepreneurs in communities has been identified as a significant gap in rural renewable energy market development. Entrepreneurs are needed to identify and develop commercial products and services based on renewable energy, to market those products and services and to repair or maintain them.

4.1.8   Others

Other noteworthy barriers to renewable energy technology transfer are:

- inadequate supporting infrastructure;
- vested interests;
- inability to pay;
- unwillingness to pay;
- lack of confidence in new technology;

- lack of access to renewable energy resources;
- technical problems.

Supporting infrastructure   This can be defined as elements that contribute to the supply and service network for renewable energy systems. It includes both physical infrastructure and human capacity. Elements of supporting infrastructure include the SMEs that are involved in the technology transfer process (e.g. selling systems and spare parts or providing micro-financing); the technical skills of people trained in installation and maintenance; the management skills for businesses (e.g. stock control); the enforcement of quality control and standards; and the transportation and communications networks for ordering and distributing technology and collecting payments.

Lack of supporting infrastructure in remote rural areas poses a barrier to the sustainable development of renewable energy systems. For example, the lack of available skilled labour for installation and maintenance of renewable energy systems and poor stock control are two factors that can lead to the failure of the technology through poor maintenance and a lack of spare parts.

Vested interests   In any society or institutional structure, there is resistance to change. One of the factors causing this resistance is vested interests. For example, there may already be long-term contracts set up for the supply of electricity from stand-alone diesel generating sets; or local electricity companies might have the sole rights to supply power to customers in a particular region, preventing independent power producers from providing power to local communities or industry. In addition, if there is continued investment in competing undesirable technology now, that technology will be used for many years until its useful life is up, causing further time delays before new, cleaner technology comes on-line.

Ability to pay   As mentioned under finance barriers, some end users find it difficult to pay for energy supplies regularly every month, as they often depend on irregular sources of income such as the harvest of crops or sale of handicrafts. Also, cash payments can be difficult to arrange in some very remote rural areas, where much trade is done in commodity form and where it would be easier for consumers to pay in commodities such as crops, livestock, handicrafts etc.

Renewable energy projects have failed in many cases due to low levels of payment collection. This is partly due to the difficulties outlined above, and partly to the difficulties involved in collecting the payments in remote areas. In many cases, particularly in remote rural areas and in peri-urban areas, end users either cannot afford to take out a loan and invest in the renewable energy systems, or are not able to secure a loan due to lack of collateral.

Willingness to pay As noted above, users may be unwilling to pay for energy supplies if they have received energy services free or at very low cost in the past. For example, if a solar home system demonstration project was provided free of charge to a village, then surrounding villages will be unwilling to pay for the services which they know have been provided free to their neighbours. Willingness to pay is therefore undermined, even though there is a demand for the services that the system can provide.

Lack of confidence in new technology Some renewable energy technology is not sufficiently proven in local conditions to be attractive to private financiers. Investors are not confident of the economic, commercial or technical viability of new technology. It is not easy for decision-makers to know when a new technology has moved from a research and technology development stage to become fully viable on a commercial scale, particularly if it is transferred to a region with significantly different local conditions.

Access to renewable energy resources Renewable energy resources are site-specific and some are also intermittent. There is often a lack of the accurate local resource data that are required to assess the real potential of a resource. There may be competing uses for some resources such as biomass, for example cooking and space heating, paper manufacture, fibreboard manufacture, packaging materials or fertilizer. Long-term contracts for biomass resources can be difficult to obtain.

Thus access to available biomass resources may be uncertain. Also, if agricultural residues start being used for power generation, they suddenly have a value and prices may increase.

Competition for land use (among agricultural, recreational, scientific or development uses) can occur among government agencies, agricultural agencies, private investors or environmental groups. This real or perceived risk can affect access to renewable energy resources associated with the land area. Also, access to resources such as small hydro may be restricted due to land ownership along the banks of the river, or extraction rights.

Technical   Four of the main reasons for technical failure are lack of understanding of local energy service requirements; lack of research and development to adapt technologies to local environmental conditions, resources and requirements; lack of local skilled labour to install, operate and maintain the equipment properly; and lack of access to spare parts.

4.2   Case studies

In order to analyse the barriers to renewable energy technology transfer to developing countries and the options for overcoming the barriers, eight case studies have been examined. The analysis is base on research undertaken in Southeast Asia; however, a few other countries have been included where the author also had information and experience. The research is based on two renewable energy technologies in particular: stand-alone solar home systems in rural areas and grid-connected biomass cogeneration. These technologies are not necessarily the most representative technologies for all developing countries, nor those with the greatest potential, but they were chosen because they cover a broad range of potential barriers and issues. These two technologies highlight many of the general barriers faced by renewable energy technologies, but obviously do not cover the technology-specific issues faced by every other technology. They cover barriers and options related to both grid-connected and off-grid systems and to access to resources. Two technologies were chosen for ease of comparing lessons learned between different countries. The full case studies can be found in Annex 1, including a summary of lessons learned in each case.

4.2.1 Summary of barriers to solar home systems

From analysing the case studies in Annex 1, it can be seen that the main barriers to successful solar home system (SHS) projects include the following:

- *Unwillingness to pay* for the systems. For example, the provision of free or very cheap energy technology from demonstration projects gives users the impression that they need pay very little or will be given the service free. Inconsistencies in fee setting make people unwilling to pay if their neighbouring village pays less. Weak action regarding system removal when households are behind on their payments encourages people not to pay, as households are still seen to be receiving the service without making payments. Social pressures on the payment collector or person responsible for disconnecting or removing systems can be difficult if they are closely related to the households with the systems. Some back-up from outside the community for monitoring and enforcement can help to neutralize the social pressures.
- *Inability to pay* for the systems. Households in rural areas do not often have regular cash incomes. Their cash income may depend on the harvesting and selling of crops, the making of handicrafts or other products, or the receipt of money from members of their family working away from home in urban areas or overseas. Some households are unable to afford systems; others can afford them but are just not able to make regular monthly payments due to irregular incomes. Also, in very remote rural areas households may trade in commodities rather than cash, so cash payments are difficult relative to payments in commodities.
- *Lack of access to credit* for all players. Access to credit for households can be a problem as there are few suitable micro-financing schemes available in rural areas (though more are being developed). Also, the interest rates charged can be high. If lending organizations

do not recognize the value of the PV modules as a loan guarantee then it can be difficult for households to secure a loan. Module suppliers could help by guaranteeing to buy back the modules at a depreciated value over time. Access to credit for other players such as developers, installers and local technology manufacturers can also be difficult, as banks are not used to assessing project finance for renewable energy systems which have a relatively high up-front cost (though training is starting to be given in renewable energy project financing in some financial institutions, e.g. the Development Bank of the Philippines). It is difficult for new companies entering the market to show they have a sound track record and are creditworthy.

- *Lack of awareness* by potential users and other key players. If potential users are not aware of or familiar with PV technology, there will be no demand for systems and no market for the technology. Marketing is a key difficulty, especially in rural areas, where access to media such as TV and radio is limited. Awareness often has to be raised face-to-face in the communities with demonstrations of the technology first-hand, which can be time-consuming but effective. If potential developers, suppliers, manufacturers, financiers etc. are not aware of the market opportunities for PV systems, they will not participate in market development.
- *Lack of government policies and incentives.* If there are no clear policies or government backing (via incentives) for the development of PV technology, then other key players, particularly the private sector, are not encouraged to invest in the technology or participate in market development.
- *Lack of confidence* in the technology. If systems are designed such that they do not meet the requirements of the people, or social and cultural considerations are not taken into account, then projects will not be successful. Also, if poor-quality technology is used or systems are not maintained regularly, then technical failure will occur. Lack of confidence in the technology will prevent financiers from providing funds and the market will be damaged.
- *Poor management and implementation.* If projects are managed poorly then they are likely to fail, as the necessary money and other resources will not be available to maintain the systems properly.

- *Lack of skills and knowledge* on the part of local people in rural areas on how to install, operate and maintain the systems leads to technical failure.
- *Lack of supporting networks and infrastructure* in remote areas. When the PV market is new and developing, there is often a lack of the supporting service networks and infrastructure required for maintenance and supply of spare parts. A delay in access to spare parts or a lack of local people trained in maintenance can lead to the systems not being operational for weeks or months, or being permanently out of action. Access to spare parts can be made more difficult with tied aid programmes that require technology to be supplied from overseas, unless local stock control can be managed effectively.
- A *lack of standards* can allow poor-quality technology into the marketplace, and users are not able to distinguish poor-quality from good-quality systems. Used properly, standards can provide good guidance to users regarding the quality and value of systems. However, if implemented inappropriately, standards can have a negative impact. For example, if the technology specification is dictated too strictly, this may make systems too expensive for users and can destroy the market.
- *Subsidies* to fossil fuels can make it difficult for PV systems to compete economically. Subsidies to grid extension programmes can also make it difficult for PV systems, as the real cost of grid extension is hidden.
- *Security.* Although this was not highlighted in the PV case studies, something that can be a problem is the theft of PV modules, as they have a good resale value and are easily transported. Batteries and other components could also be stolen, but are more easily locked away in buildings, or watched over by the owners. The issue of security has been addressed in the Shell–ESKOM joint venture in South Africa. Each component of the solar home systems that they install (e.g. the battery, controller, PV panel etc.) is electronically coded. They are then connected to a central unit called the 'powerhouse', which houses the battery and controller. Each component connected to the powerhouse is initialized before the system can

work. After that, if at any point the components are removed they will not work connected to any other system. Even if the whole system is stolen it will not work, as the powerhouse has an anti-tamper device which is set off if moved, after which the system has to be reset with a special code.

4.2.2   Summary of barriers to biomass cogeneration

From analysing the case studies in Annex 1, it can be seen that the main barriers to successful biomass cogeneration projects include the following:

- *The financial crisis* in 1997 brought a halt to small power producer (SPP) negotiations in most Southeast Asian countries, but these negotiations began to resume in 2000.
- *There is a lack of knowledge and awareness* among mill owners and potential developers in some countries regarding the technical and economic case to support efficient biomass cogeneration.
- *Lack of awareness among potential developers and mill owners* of the support packages on offer from central government, or of ODA from bilateral or multilateral institutions.
- *Lack of a successful commercial track record* and experience in efficient cogeneration pilot projects for some biomass sources.
- *Lack of awareness and skills regarding the need to adapt technology to local conditions.*
- *It can be difficult to make sure operation and maintenance guarantees are upheld.*
- *The high pressure cogeneration systems are seen by mill owners as complicated to run*, as power generation is not their core business.
- *Unfavourable conditions associated with power purchase agreements*, deterring SPPs from setting up and reducing the confidence of potential investors.
- *Problems of fuel supply.* Security of fuel supply is a continuing risk as the agro-industries are not geared up to long-term contracts. Once waste starts being used for power generation it is seen to be of value, so prices may rise. Seasonality of fuel supply results in non-firm

power generation (unpredictable, variable levels of power output) and difficulties in securing PPAs or good power purchase prices.

- *Problems of resource supply.* If a generation plant is not integrated into a mill and the mill owner has no incentive to supply waste streams to the cogeneration plant, then there is little guarantee of biomass supply to the power plant, particularly as there is growing competition in other uses for agro-industry wastes. It is best if the biomass fuel supplier is bound to the project on long-term delivery contracts, or is made part of the power enterprise or consortium.
- *Poor relations between the developer and mill owner* may prevent commercial information being kept in confidence while investment is sought.
- *Lack of supporting institutions,* which do not yet exist in some countries, and of relevant specialist advice.
- *Lack of clear government plans and targets* regarding renewable energy; this sends poor signals to potential investors and developers.
- *Lack of confidence on the part of investors.* The impacts of mill size, fuel transportation and seasonality of fuel supply/electricity demand are difficult to quantify; thus investors are apprehensive and financial barriers are created, particularly high-risk premiums set by financiers who do not understand the technology or sector.
- *Weak and unstable institutions* involved may cause projects to be unsustainable or increase the perceived risk to financial organizations.
- *Foreign exchange risks* are very real, as was made clear by the economic crisis in Asia in the late 1990s. These can be minimized for the developer if power payments are linked to the US dollar. However, the currency risk is then heightened for the utility, which would be hit by the drop in exchange rates. There is not likely to be a full transfer of risk, but maybe a percentage of the payment could be linked to the US dollar to spread the risk.
- *Lack of awareness of how to access loans and grants* for pre-feasibility studies and feasibility studies, which are needed to develop the market in the early stages and help to reduce the perceived investment risks.
- *Subsidized electricity supply systems* make it difficult for renewable energy projects to compete economically.

- *Restrictions on third party sale*, which if allowed would enhance the viability of projects, providing shorter payback periods, and hence accelerate investment in more cogeneration capacity.

4.3 Options

4.3.1 Encouraging technology transfer

In order to achieve successful, sustainable transfer of renewable energy technologies to developing countries there are some issues that need to be addressed to encourage the transfer process.

First of all there needs to be recognition that the requirements and demands for energy services are different in developing countries from those in industrialized countries. There are three main types of demand for energy services in developing countries: (1) grid-connected electricity in large towns and cities; (2) electricity for medium-sized settlements via mini-grids; and (3) a mix of thermal, mechanical and electrical power from stand-alone systems mainly in remote rural areas. This contrasts sharply with industrialized countries, which are almost entirely grid-connected and in some cases also have extensive gas mains access.

Second, the role that renewable energy can play in fulfilling energy service requirements, and the benefits that renewable energy can bring, need to be recognized and valued. For example, the contribution renewable energy can make towards meeting development targets and local, regional and global environmental improvements should be appreciated. In acknowledgment of this, renewables should be integrated into development plans and programmes alongside other energy options and energy efficiency objectives.

Third, to secure a market for renewables and plan effectively to meet energy service demands, long-term energy plans, including specific targets for renewable energy, are needed to underpin market development and create confidence in the market. Many developing countries already have long-term energy plans and an increasing number are developing specific targets for renewable energy. In addition, industrialized countries can lead the way with renewable energy uptake to show that it is not a second-class option for energy provision. Deploying renewable energy technologies in industrialized countries will increase manufacturing volumes and help to bring costs down, making the technology more affordable for developing countries.[6] Where renewables are not already competitive with fossil-fuel technologies, the development of strong global markets will help them become more competitive.

Fourth, existing policies and regulations need to be looked at and understood before considering changes that create incentives for renewables. Hidden subsidies need to be identified and made transparent in order to level the playing field (e.g. by removing or reducing subsidies on fossil fuels).

Fifth, the number and capacity of local SMEs must be expanded if they are to support the marketing, sale, installation, operation and maintenance of an expanding number of renewable energy systems. The technical and business skills of local staff should also be increased to help manage the SMEs and implement projects and systems.

Sixth, collaboration among countries and organizations is essential in order to help facilitate the transfer of technology and skills. Partnerships can help develop manufacturing capabilities locally, and joint R&D – particularly between developing countries – can speed up the adaptation of technologies to local conditions.

Seventh, it is essential that we learn from past experience and analyse previous and current projects to identify the lessons to be learned and to start to identify best practice for project and programme implementation. Successful projects need to be replicated using best practice in order to speed up the deployment of renewable energy technology.

Finally, issues regarding access to finance and best use of funds need to be understood. The constraint on financing is not caused by a shortage of investment funds, but rather by an inability to attract available funds due to a lack of confidence in and understanding of renewable energy markets and technologies. New appraisal methodologies for renewable energy systems are being developed to help investors understand the investment risks and investment profile of renewables projects. Micro-financing schemes are being tried out and developed to match the borrowing requirements of households and small-scale

---

[6] G8 RETF (2001).

organizations. Developers need to be given the necessary training to help them put together 'bankable' projects.

Two essential things that need to happen early on to establish a solid base for renewable energy technology transfer are building confidence in a wide range of actors and creating markets. Confidence must continue to be built in the technologies (reliability), in projects (profitability and sustainability) and in the market for energy services. There is a range of different actors who need to have confidence in renewables in order to facilitate the transfer process and the development of markets. Demonstration projects along with dissemination and awareness-raising programmes can build confidence in the technology among a range of key actors including potential investors, developers and end users. Commercial-scale pilot projects are extremely valuable in showing potential investors that they can get a return on investment. Government plans and targets for renewable energy can help build confidence in the potential market for renewable energy services, thus encouraging developers to get involved in providing the required services. Regulations regarding independent power producers and the terms and conditions of PPAs can be developed to encourage and secure the market for electricity from renewable energy. For example, standardized PPAs and 'must buy' contracts or priority despatch for renewables are needed, and the matter of firm versus non-firm power production needs to be addressed. Pricing of renewable energy projects needs to be competitive; this may require fiscal incentives or minimum purchase requirements, such as the renewable energy obligation in the United Kingdom (which has superseded the old non-fossil fuel obligation). This reduces the perceived risk of investment and can therefore increase the confidence of potential investors that there will be a market for the power for long enough to ensure them a payback on the investment cost.

In order to create markets, the energy service requirements of potential customers need to be understood and technology adapted to local conditions and demand patterns. Once potential users see the technology in use and the benefits it can bring, a demand is created which is essential for market development. Market and resource assessments are good tools with which to determine the potential of different technologies.

In order for markets to grow in a sustainable way, the energy services provided must be affordable and cost-competitive with other energy sources. Some renewable energy technologies are commercially competitive with other forms of energy, but in most cases costs still need to be brought down in order to enable them to compete in the market. This can be done through continual R&D to improve designs and manufacturing techniques, and through learning effects. If manufacture of the technology can be stepped up, cost savings will result from economies of scale. Local manufacture can also bring down costs by virtue of lower labour cost and raw material costs; however, this will only be cost-competitive if the scale of manufacture is large enough. In some cases components of the technology are too high-tech to manufacture locally, but there are usually some components that can be made in-country (e.g. blades and towers for wind turbines; PV panels in which to mount the imported PV cells). Removal or reduction of subsidies and other assistance given to fossil fuels can help renewables compete on a more even basis. Fiscal policies can be used to help make renewables more affordable (e.g. by removing import duty and tax on renewable energy components). Where renewables are seen as essential to development and poverty reduction, 'smart' subsidies can be introduced that target a specific requirement, benefit the poor, are transparent and can be phased out over a period of time without destroying the market. If the full costs of fossil-fuel externalities are taken into account and the benefits created by renewables are valued, then the life-cycle cost of energy services from renewable energy would in many cases show them to be the cheapest and most beneficial option.

Other things that can be done to reduce the cost of renewables and make them more affordable include reducing transaction costs by streamlining the planning and development process ('one stop shops' for planning and contracting) or bundling projects into groups for investment, thus spreading the transaction costs. However, projects need to be chosen very carefully for bundling, as a delay in one project will delay investment in all projects.

It is extremely important to create appropriate incentives to attract relevant actors to get involved in the planning, implementation, development and use of renewable energy systems. Incentives are needed to attract investment and encourage partnerships (e.g. tax holidays). A strong legal and regulatory structure is also needed to encourage investment and partnerships, including intellectual property rights, and setting and enforcing appropriate technology standards.

Also important in assisting the technology transfer process, both at an early stage and once it is under way, is the exchange of up-to-date information between industrialized countries and developing countries, and between developing countries. This can be done via networks or partnerships and is important for raising awareness and enabling all parties to make informed decisions.

4.3.2 Overcoming the barriers in developing countries

It is very difficult to pinpoint the specific conditions required for best practice when looking at overcoming the barriers to renewable energy technology, as each situation can have many local, national and international influences on the successful outcome of a project. Some of these factors can have a direct and very obvious impact – for example, lack of access to credit. Other factors can have an indirect impact – for example, a previous aid-funded project may have provided electricity free of charge to nearby villages, so that people are now unwilling to pay for electricity even though they may value it highly and there may be a market for its use.

There is no 'one fits all' solution to successful technology transfer. The identification, analysis and prioritization of barriers needs to be done on a country-by-country, case-by-case basis. It is important to identify the most appropriate actions and policies that address the specific barriers faced in any one situation and that give maximum rewards to the actors involved by addressing their interests and requirements; a balance has to be found between the requirements of the recipients of the technology and those of the instigators and investors in the transfer process.

Although there are no all-inclusive solutions, some lessons can be drawn out of both successful and unsuccessful projects. If these lessons are considered when implementing new projects this can increase the chances of success, but will not guarantee it, as factors specific to that region may impact in ways that may not be predictable or obvious at the time of implementation. (It is always much easier in hindsight to see why projects have not worked out!)

Some of the barriers to renewable energy technology transfer in developing countries are similar to those faced by renewable energy market penetration in industrialized countries, so lessons can be learned from these experiences. A study conducted for the IEA on the barriers to renewable energy market penetration in industrialized countries identified a chain of support measures for renewable energy technologies designed to address barriers at different stages in the process of market penetration.[7] Figure 4.2 shows the chain of support measures, adapted to take account of capacity barriers found in developing countries.

As technology progresses along the path of market penetration in industrialized countries (from the top to the bottom of the figure) it encounters different barriers, as shown down the left-hand side of the figure. Support measures that can be taken to overcome these barriers are shown in the boxes. This chain of support image can apply to renewable energy technology transfer to developing countries. However, there are some measures which are not so relevant to developing countries which lack the appropriate technical capacity, skills and investment. For example, investment in high-tech R&D should mainly be the role of countries with the funds, skills and technical capacity to do so, i.e. industrialized countries. Nevertheless, some larger developing countries, such as China and India, also have pockets of skills and money for research activities, particularly as labour and materials are cheaper than in industrialized countries. In some cases, actions taken by developing countries can be supported with cooperation from industrialized countries; these might include training and capacity building, resource assessments and setting of standards.

Annex 2 contains two tables showing suggested options for overcoming the barriers identified in the case studies in Annex 1. These should be regarded as guidelines only, as unique local and national situations regarding policy, legislation, environment and culture will

---

[7] Smith and Marsh (1997).

Figure 4.2: The chain of support for renewable energy technologies

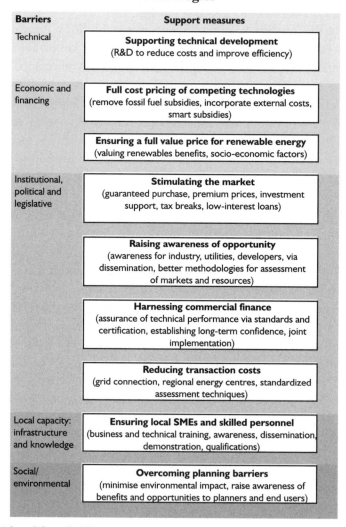

Barriers	Support measures
Technical	**Supporting technical development** (R&D to reduce costs and improve efficiency)
Economic and financing	**Full cost pricing of competing technologies** (remove fossil fuel subsidies, incorporate external costs, smart subsidies)
	Ensuring a full value price for renewable energy (valuing renewables benefits, socio-economic factors)
Institutional, political and legislative	**Stimulating the market** (guaranteed purchase, premium prices, investment support, tax breaks, low-interest loans)
	Raising awareness of opportunity (awareness for industry, utilities, developers, via dissemination, better methodologies for assessment of markets and resources)
	Harnessing commercial finance (assurance of technical performance via standards and certification, establishing long-term confidence, joint implementation)
	Reducing transaction costs (grid connection, regional energy centres, standardized assessment techniques)
Local capacity: infrastructure and knowledge	**Ensuring local SMEs and skilled personnel** (business and technical training, awareness, dissemination, demonstration, qualifications)
Social/ environmental	**Overcoming planning barriers** (minimise environmental impact, raise awareness of benefits and opportunities to planners and end users)

Source: Adapted from Smith and Marsh (1997).

impact greatly on the success of projects. The options are summarized in Table 4.2. The table also presents the impacts that options can have, showing in some cases options addressing more than one set of barriers.

Table 4.2: Options for overcoming barriers

Option	Impact
National policies and programmes	
Set long-term (5–10-year) energy and electrification plans with clear targets for renewable energy	Increases confidence of potential investors in renewable energy
Liberalization and privatization of electricity utilities, allowing IPPs and SPPs to sell electricity to the grid	Allows electricity from renewables projects to be exported to the grid on favourable terms, encouraging investment in renewable energy power projects
Standardization of PPAs including 'must buy' clauses, minimum renewables obligations and favourable pricing	Provides more favourable conditions and competitive prices for renewables to export power to the grid
Fiscal policies to provide incentives for renewable energy, including reductions on tax and import duty and smart subsidies where there is a social need	Helps to make renewables more cost competitive with fossil fuels, assisting the poorest people to afford renewable energy services for development and poverty reduction
Pilot and demonstration schemes, including full-scale commercial projects	Checks the technology is effective in the local conditions and meets the required demand
	Raises awareness among actors, encouraging demand and market development
	Increases investors' confidence in technical and economic viability
Institutional structures	
Cooperation and coordination on renewables between government departments; integration of renewables into development plans	Helps achieve development goals and raises awareness of the role renewable energy can play; improves understanding of energy requirements for different sectors (health, education, transport, communications etc.)
'One-stop-shop' for energy planning	Reduces duration of planning procedures and transaction costs
A 'champion' for renewable energy	Identifies one department with the responsibility and drive to meet the targets
Intellectual property and standards	
Clear IPR and strong legal institutions	Increases confidence for investment in joint ventures, joint R&D and local manufacture of renewable energy technology and direct current appliances

Table 4.2: continued

Option	Impact
Appropriate technology standards and strong enforcement	Allows informed choices against both cost and performance
	Increases confidence in technology for end users and investors
Information exchange, education and training	
Feasibility studies, resource assessments and market assessments	Provide information with which energy services can be planned
Set up information networks and centres of excellence for collection, analysis and dissemination of information via appropriate media	Learning lessons from existing projects and identifying best practice
	Raising awareness of renewable energy services among actors
	Exchanging information between developing countries as well as between industrialized and developing countries
	Allows people to make informed decisions regarding energy services.
Education and training, in technical and business management skills. Courses in schools and universities, training of women as well as men.	Local technically skilled staff
	Local business management staff that are able to help increase the number of and expand the size of SMEs
	Training women can help to keep the skills local, as men are more likely to move to bigger towns or cities with new found skills
Financing	
Loan guarantees and recognition of the value of renewable energy systems as an asset against a loan (particularly PV panels)	Reduces risks for investors and facilitates access to finance
Renewable energy project appraisal methodology, accounting for small size of projects, and high up front costs	Renewable energy projects appraised more fairly and investor understanding increased, reducing perceived risks
Life-cycle costing of all energy projects including externalities and benefits	Externalities of fossil fuels would increase their relative costs and the benefits of renewables would reduce their relative costs, making renewables more cost competitive.
Grants and soft loans for: pilot, demonstration and full scale commercial projects; SMEs to develop a track record and gain credibility with commercial banks; skills and knowledge transfer	Increasing confidence in technology for investors
	Local capacity building and credibility

Table 4.2: continued

Option	Impact
Provision of high risk loans by development banks, bilateral and multilateral aid agencies for SME development	SME access to loans when commercial banks see high risk Building up the track record of SME operation reduces investment risks and attracts investors
Revolving funds and innovative finance mechanisms (e.g. SDG, AREED, REEF, CDM, 'patient capital')	Provide a range of finance options to suit the local requirements Investment for technology and skills transfer and capacity building
Micro-financing schemes with local organizations	Small businesses and households can get access to credit Better local management of loans and fee collection
Bundling of small projects	Can reduce transaction costs
Smart subsidies	Targeted, pro-poor and transparent subsidies reaching those in need Subsidies that can be removed without destroying the market created
Social	
Participatory planning	Meets the requirements of local people; gains social acceptance for new technology
Involve entrepreneurs and agents for change	Drive projects and programmes and catalyse action
Clear payment collection and strict disconnection policies with appropriate collection frequency and method of payment	Improves payment collection levels and the economics of projects, increasing investors' confidence Matches income patterns with payment dates and considers non-cash payments in rural economies
Use respected local organizations for payment collection	Social pressure increases prompt payment
Use energy service companies providing a fee for service	Allows poor people who cannot afford to purchase systems access to energy services
Other	
Encourage partnerships and joint ventures	Secures access to renewable energy resources such as biomass Enables local manufacturing facilities to be set up, bringing down technology costs

Table 4.2: continued

Option	Impact
Encourage research and development	Facilitates adaptation of technology to local conditions and requirements, bringing down technology costs
Raise awareness of business opportunities and job creation from renewables	Can reduce inertia from those with vested interests in competing energy services if they see new opportunities

Options regarding national policies and programmes   Setting long term (5–10-year) energy and electrification plans with clear targets for renewable energy that are adhered to will increase confidence of potential investors by showing that there will be a long-term market for renewable energy.

Liberalization and privatization of electricity utilities can open the way for independent power producers (IPPs) and small power producers (SPPs) to sell electricity to the grid. Terms and conditions of power purchase agreements (PPAs) should be set to encourage renewable energy (i.e. 'must buy' contracts, no penalization for variable power production, etc.). This allows electricity from renewables projects to export power to the grid and encourages investment in them.

Fiscal policies can provide incentives for renewable energy (e.g. by removing tax or import duty on renewable energy components, or offering tax holidays for new renewable energy companies) and level the playing field for renewables (e.g. by removing subsidies on fossil fuels, or introducing sustainable targeted 'smart' subsidies for renewables where there is a social or developmental need). This helps to make renewables more cost-competitive with fossil fuels and helps the poorest people to afford renewable energy services for development and poverty reduction. National governments need to have clear policies on subsidies so that developers can assess the costs of implementation. They also need to have straightforward mechanisms to implement the subsidies so that eligible developers and/or end users are not deterred from applying for them.

Pilot and demonstration projects are important as they can be an efficient way of checking that the technology is effective in the local conditions and meets the required demand, building confidence among actors. They can also raise awareness about the technology among policy-makers, installers and end users, encouraging demand and market development. Full commercial-scale demonstrations can show the economic viability of projects and thus reduce the perceived risk of investment.

Options regarding information exchange, education and training
Feasibility studies and resource and market assessments are the first step in determining what potential there is for a technology. These need to be accurate and detailed assessments that should cover the whole country, but specifically those areas where there is a demand for energy. Market assessments are important for determining the potential demand for energy services.

Many different technologies are being demonstrated in developing countries. Some projects are successful and others fail for particular reasons. It is important for this information to be shared among countries so that all can learn from others' experiences and not make the same mistakes. It is important to pool information on projects so that the experiences can be analysed and common barriers and critical success factors identified. Replicability is important if renewable energy technology is to be transferred on a large scale. Centres of expertise can assist the collection and analysis of information. Information networks can play an important role in helping to disseminate information and raise awareness. Awareness raising is an important step in the technology transfer process, and it is important to set up networks to exchange information among developing countries as well as between industrialized and developing countries. It is important that the information provided to potential end users, installers and developers is relevant, and is presented in a form that is easily accessible and familiar to the target audience. For example, many people in rural areas are not literate, so printed information would not be of much use to them. A more appropriate vehicle might be the radio, or verbal sharing of information at community meetings. Providing information and raising awareness can help people to make informed decisions and create demand for energy services.

It is very important to build a local capacity of technically trained staff to help install, operate and maintain the equipment. It is particularly important to have local skilled labour in remote rural areas where access is difficult and transport infrequent. After the technology has been in place for a few years, the local technical staff need to become more skilled in troubleshooting as problems begin to occur and replacement parts are needed. Equally important is training in business management skills to help increase the number and size of SMEs required for the retail, installation, operation and maintenance of systems.

Options regarding institutional structures   Institutional capacity building is needed to provide developing countries with the skills and structures to assist with the planning, implementation and development of renewable energy systems. Cooperation and coordination on renewables between government departments is important to assist the integration of renewable energy in development planning and across all relevant sectors – such as health, education, transport, communications, agriculture, small industry, etc.

A one-stop shop for energy planning can speed up the process and benefit renewable energy projects by reducing transaction costs. Transaction costs are more dependent on the duration of project negotiations than on the size of the project, so they can be prohibitive for small-scale projects if planning and negotiations are long-drawn-out.

Targets for renewable energy will be met only if there is a single person or government department charged with delivering those targets (a 'champion' for renewable energy). The performance and success of this department need to be measured by their ability to meet the targets; otherwise there is no motivation for achieving them.

Options regarding intellectual property and standards   Clear intellectual property rights and strong legal institutions to enforce legal judgments are needed to encourage foreign private companies to enter into joint ventures with local companies to manufacture technology locally or to carry out joint R&D. In some cases, where weak IPR are deterring joint ventures, some of the less high-tech components of renewable energy systems are beginning to be manufactured locally.

These include towers for mounting wind turbines, batteries for use in solar home systems and encapsulation of photovoltaic cells into panels.

Appropriate and harmonized standards, regulations and guidelines enforced at regional and local levels can help alleviate perceived risks regarding quality of technologies. For example, in Kenya it has been known for a PV panel to be sold with a colour photocopy of PV cells in it rather than the real thing; a quality standards marking can help in guarding against such fraudulent practices. Standards also allow people to make informed technology choices on a cost as well as performance basis.

Options regarding financing   Loan guarantees can help secure access to finance. These can be provided by a range of actors including local government in host countries, developers and technology suppliers, or industrialized countries via their export credit agencies (ECAs). Currently ECAs provide risk insurance and investment for exports of a range of technologies from industrialized countries. Renewable energy technologies do not benefit much from this at present as the ECAs are not familiar with the technology and are therefore reluctant to invest in it. ECAs and other potential investors need to be trained and educated in project appraisal methodology suitable for small renewable energy projects, taking account of their high up-front investment profile. They need to be aware of how to carry out due diligence assessments for potential projects, which assess the risks and feasibility of the project; they also need to be aware of the benefits of renewable energy and the externalities of fossil fuels. The fairest and most equitable way of comparing the cost of different energy systems is by carrying out a life-cycle analysis, covering all costs and subsidies, and including costs for the externalities and benefits of each system. In this way the costs from fossil-fuel externalities and the benefits from renewables will show renewables to be the least-cost option already. Insurance companies are starting to realize that investments in fossil-fuel power plant are indirectly contributing to their higher insurance payouts caused by extreme weather conditions, which are in part due to global warming as a result of carbon dioxide emissions from fossil-fuel power plant. They are therefore in effect shooting themselves in the foot by investing in fossil-fuel power plant and, having acknowledged this, are starting to consider clean energy options for investment.

National investment plays an important role in financing feasibility studies, pilot and demonstration projects on a small scale, and investing in training and education required for capacity building. Foreign investment can play an important role in providing funds for studies and projects on a larger scale and also in increasing the number of systems in a country so that economies of scale can be gained, making the systems more affordable and sustainable. Investment in fully commercial demonstration projects can raise the confidence of potential investors. Investment in R&D is important to help bring down technology costs and adapt technology to local conditions and requirements. Investment in training, education and business management is also important to transfer skill and know-how. It is important that loans from development banks or multilateral and bilateral aid agencies that are willing to invest in higher-risk developmental projects are made available to SMEs to help them build up a track record which will give them credibility with commercial banks.

Revolving funds and other innovative finance mechanisms are being developed for renewable energy project financing. Some of these are outlined in Chapter 3; they include REEF, SDG, CDM and 'patient capital'. Some are already operational, but others are still in the planning stage; for example, at the time of writing, the details of the CDM are still under negotiation.

Access to finance for rural households and small businesses can be addressed through micro-financing schemes and tailored to local requirements. There is a need to develop more micro-financing schemes and expand existing ones geographically to increase access to appropriate finance.

Transaction costs can be high in proportion to the cost of small renewables projects, so bundling of projects can assisting in spreading the transaction costs among several projects, bringing down the cost to each.

Rural energy systems all over the world are subsidized (in both industrialized and developing countries). Subsidies are often used to reduce the cost to the end user. It is important when implementing subsidies to ensure that the subsidy is sustainable, or that when the subsidy is reduced to a sustainable level the market is not destroyed. It is also important to make sure that subsidies benefit those they are intended to help. Carefully targeted 'smart subsidies' can be implemented that more effectively benefit the poor. Subsidized fossil-fuel energy prices deter investment in alternatives. Different forms of levy to 'internalize' the cost of fossil-fuel use are now being tried in some countries to improve the competitiveness of the cleaner energy sources.

Options regarding social issues Community involvement in the planning and implementation (participatory planning) of renewable energy projects in rural areas is very important, as the technology must be chosen and designed to meet the requirements of the local people and to suit the local conditions. Involving the recipients in the choice of technology also helps to establish social acceptance of it.

Poor payment collection is often a problem in rural energy projects. If strict action is not taken to cut off those end users who do not keep up payments, then other end users who are paying see there is no need to do so and also stop paying. It is therefore very important to have both a clear payment collection system and strict disconnection policies. The choice of both the organization and the individuals in charge of collecting payment is important. They need to be well respected, as social pressures can be strong in rural communities and can exert influence both ways, deterring payment collection if the organization or payment collector is not well respected, or on the other hand encouraging prompt payment if the household is frowned upon by other members of the community if they are behind on payments. A good example of applying social pressure to gain prompt payment is the Vietnam Women's Union (see Case Study 3 in Annex 1).

Method of payment and frequency of collection need to suit local circumstances. For example, if a community is reliant on the sale of crops for a cash income, then the collection of payments needs to match the seasonal pattern of crop harvest. In very remote communities, commodities are often traded with each other rather than for cash. This also needs to be taken into account when agreeing the method of payment.

It may be feasible to take payment in labour or commodities rather than in cash. Value for money in local comparative terms must be understood and addressed by developers and suppliers.

In many cases users just cannot afford to take out a loan to purchase systems. One way around this is to set up a rural energy service company (RESCO) that remains the owner of the technology. The end user would then pay for the services the RESCO can offer (i.e. the supply of electricity or bottled gas). In this way, the end users could be asked to pay a connection fee (just as they would for grid electricity) and then a monthly (or more appropriately timed) fee for the service they receive. In the South Pacific island state of Kiribati, for example, aid funding provided the initial solar home system technology and financed the solar energy company to install the systems. The monthly payments cover the salaries of the head office and field maintenance staff, and the replacement of batteries, controllers, wiring and switches. The end users pay for replacement lights. If the number of systems reaches a critical mass (around 1,000), the payments will cover the replacement of solar panels as well, and the energy service company becomes sustainable.

Options regarding other issues   It is important to establish partnerships and joint ventures between local and foreign organizations. This can help develop local manufacturing of technology at a lower cost than imported technology. It can also assist in securing a supply of biomass for power generation if the mill owner or biomass supplier is made a partner in the project, with benefits following if the project is successful.

R&D is important in order to adapt technology to local conditions and requirements, and to bring down technology costs.

There is a certain amount of inertia in the adoption of renewable energy technologies due to vested interests in other technologies. This can be addressed to some extent by raising awareness on the part of utilities and other organizations of the new business opportunities offered by renewables and encouraging them to look at renewable energy as a complement to existing energy services rather than a threat, particularly in off-grid rural areas and where embedded local generation can strengthen the grid. Inertia in the uptake of technology is also due to lack of knowledge and awareness among potential users. It is important to stimulate the desire of potential users to have the technology (market pull), in combination with the supply of new technologies (market push). Ethical marketing and customer communications have a critical role to play in developing market pull.

Summary From the tables in Annex 2 and the summary in Table 4.2 above, it can be seen that most of the options aim to achieve one or more of the following things: bringing down the costs of renewables and making them affordable; increasing confidence in renewables; attracting investment into renewables; building local capacity; developing supporting infrastructure (e.g. SMEs, legal and regulatory systems, transport and communications); increasing awareness to stimulate demand and market development.

Chapter 5

The Way Forward

5.1 Actions needed

Chapter 4 identified options that can be taken to address certain barriers to renewable energy. In most cases action needs to be taken by a combination of actors in order that projects and programmes are successful and barriers overcome. This chapter summarizes the specific actions highlighted by the case studies described in Annex 1 and more common actions that are needed to ensure the successful transfer of renewable energy technologies in general. It also looks at which actors could be engaged in different actions, and which actions are least politically sensitive and most likely to be carried out.

5.1.1 Actions needed for solar home systems

It has been identified in the case studies that the cost of SHS needs to be brought down so they are more competitively priced. Removal of import tax on PV technology can help to lower the cost to consumers, making the systems more affordable to buy. Costs can be reduced further through continued research and development, learning effects and increasing the volumes manufactured to create economies of scale. Smart subsidies can also be considered that target specific beneficiaries, namely the poor, and are time-limited and transparent.

To encourage investment in SHS projects, financial institutions need training in project finance methods suited to renewable energy systems. Governments can set clear rural electrification plans and targets for PV to help attract investment. To assist access to credit, micro-financing schemes need to be made available, taking into account seasonality of income and the need for non-cash methods of payment. Other schemes that allow consumers access to the technology if they can not afford to purchase it outright include leasing and RESCOs providing a fee-for-service. If technology suppliers can guarantee buyback of equipment at a depreciated value over time, this could be considered by finance institutions as a type of loan guarantee and thus help facilitate access to credit for end users.

It is important to help consumers weigh up the cost and performance of different systems, as in many cases they will have limited funds to purchase the technology and want to make sure they get best value for money. Clear standards and certification can act as a guide, helping consumers to make informed decisions. In addition, energy labelling can also provide consumers with information to help them select the most appropriate technology for them.

Local capacity needs to be developed to support the deployment of systems. The skill base can be developed through school and university courses on installation, maintenance, marketing, financing and business management. Local skills can also be developed with the implementation of joint ventures that set up local manufacturing facilities. Actions need to be taken to develop the skills of local entrepreneurs, enabling them to identify and develop commercial PV products and services, to market the products and services, and to maintain or repair them.

Development goals and policies need to be in place to ensure projects are appropriate for local requirements. Action needs to be taken to ensure that the project developers consult closely with the communities that are intended to receive the technology. The local culture and energy requirements need to be understood, and community representatives need to be consulted during participatory planning. This will help SHS to be appropriately designed, and understood and desired by the users. Key local staff and organizations need to be selected carefully so as to neutralize social pressures on, for example, payment collectors and system disconnectors and, where appropriate, to exert social pressure on those who fall behind on making payments. The complete energy needs of the household need to be taken into consideration in order to create sustainable energy solutions. Therefore, SHS need to be considered as part of a larger energy plan for households, addressing all their energy requirements (e.g. cooking, lighting, space heating) with an appropriate range of fuels and technologies. Environmental impacts need to be considered and appropriate facilities provided for battery recycling, including disposal or reuse of electrolyte, lead and plastic. Deposits on batteries can encourage the return of old batteries to recycling centres.

These actions, although drawn out of the case studies discussed in Annex 1, are relevant for a range of SHS projects and other off-grid renewable energy projects.

5.1.2   Actions needed for biomass cogeneration

It has been identified in the biomass cogeneration case studies that awareness needs to be raised among mill owners and other actors of the environmental and economic benefits of using crop residues and other biomass waste for electricity and heat production. Reducing the cost of imported technology by removing import tax can be an important sign of support for project developers from the host government, encouraging them to invest time and money in the host country to help develop biomass cogeneration opportunities. Privatization and liberalization of the electricity supply industry can allow IPPs to export electricity to the grid, thus creating a market for the electricity they produce. To encourage mill owners to export power to the grid, favourable terms and conditions need to be created in power producer agreements (PPAs), for example, 'must-buy' clauses. Tariff rates may need revision so as not to penalize small renewable energy power producers for variable load outputs. Other incentive schemes can be put in place to encourage mill owners to generate and export power to the grid, for example by allowing special payments to renewable energy generation for up to 5–15 years, guaranteeing the market over the payback period. This encourages investment in the projects and allows the mill owners or developers to secure loans. In addition, financial institutions need to be trained in project finance methods suited for renewable energy systems.

To assist the development of biomass cogeneration, there needs to be capacity building and local skill base development through school and university courses. Both technical and business management skills need to be increased locally to help develop the SMEs that can plan and select both the most appropriate systems and the supporting infrastructure needed to install, maintain and operate them. Clear standards and certification can help guide consumer choice and decision-making. More local entrepreneurs are needed to identify and develop commercial biomass cogeneration projects and to market the systems. Partnerships can help secure investment, increase access to know-how and increase biomass resource security.

These actions, although drawn out of the case studies discussed in Annex 1, are relevant for a range of biomass cogeneration projects and other grid-connected renewable energy projects.

5.1.3 Common actions needed

Apart from the specific actions identified above for solar home systems and biomass cogeneration projects, there are some common actions that are needed to support the transfer of renewable energy technologies in general. These are outlined in Table 5.1.

As discussed in Chapter 2, there is a range of key actors that influence renewable energy technology transfer. These can be grouped as follows:

- *developing country government* (national and local), including policy-makers, legal and regulatory bodies, financiers, utilities and academic institutions;
- *industrialized country government*, including bilateral aid agencies and export credit agencies;
- *IFIs*: multilateral institutions including the World Bank, GEF, UN agencies and regional development banks;
- *private companies* including developers, installers, manufacturers, financiers, consultants, privatized utilities, and technology and biomass fuel suppliers;
- *civil society*: NGOs, communities and individuals in host countries.

Each actor has an influence on the success of renewable energy projects and the transfer of technology. Table 5.1 shows which actors play the main role in different actions and which actors play a supporting role.

5.2 Engagement of key actors

As can be seen from Table 5.1, many actions need involvement from both the public and the private sector, so public–private cooperation

Table 5.1: Common actions needed to overcome the barriers to renewable energy technology transfer in developing countries

Common actions needed	Developing country government	Industrial-ized country government	IFIs	Private compa-nies	Civil society	Political sensitivity
National policies and programmes						
Setting clear RE policies and targets	●	○	○		○	H
Removing/reducing import duty and tax on RE technology and DC appliances	●					M
Creation of support mechanisms to help fund feasibility studies and demonstration projects	●	○	○	○		M
Clear and firm plans, targets and policies for rural electrification	●	○	○		○	M
Integrated planning (energy and development), including more inter-disciplinary work between ministries (energy and e.g. health, education, water, rural development etc.)	●	○	●	○	○	M
Revise planning regulations to speed up RE applications	●	○	○		○	M
Institutional capacity building for energy planning, policy-making, regulation, etc.	●	○	○			L
Create an RE government department or other body responsible for coordinating RE planning and development: a 'champion' for RE	●	○	○			M
Remove/reduce subsidies to fossil fuels and grid electricity	●	●	○			H
Electricity supply restructuring, regulation and ownership (cost transparency, PPAs and supply to third parties)	●	○	○			H
Support and promotion of CDM and inclusion of RE on the list of priority projects	●	●	●	○	○	H

Table 5.1: continued

Common actions needed	Developing country government	Industrialized country government	IFIs	Private companies	Civil society	Political sensitivity
Establish a transparent, participatory, accountable and pro-poor regulatory framework	●	○	●	○	○	H
Intellectual property and standards						
Strengthen legal institutions	●	○	○			M
Strengthen national law on intellectual property rights and comply with the TRIP agreement	●	○	○	○	○	H
Set and enforce standards for RE technology and DC appliances, as a guideline for users, installers and developers	●	○	○	●	○	M
Develop and disseminate standard certification methodologies for resource assessment	○	●	○	○	○	M
Information exchange, education and training						
Resource and market assessments	●	○	●	○	○	L
Pre-feasibility and feasibility studies	●	○	●	●	○	L
Collection, analysis and dissemination of information regarding RE, by a central department/organization designated to 'champion' RE	●	○	○	○		L
Collaboration and cooperation between existing RE-related information networks nationally and internationally, with exchange of experience and information between South and South as well as North and South	●	●	●	○	○	L
Dissemination of lessons learned from existing projects and best practice	●	●	●	○	○	M
Raising awareness among developing country governments, development cooperation policy-makers and multilaterals of the role RE has in meeting IDTs	●	●	●	○	○	L

H = high; M = medium; L = low. ● Main action. ○ Supporting action, assistance.

Table 5.1: continued

Common actions needed	Developing country government	Industrialized country government	IFIs	Private companies	Civil society	Political sensitivity
Raising awareness of industry to the potential RE markets in developing countries, including niche markets	○	●	○	○	○	L
Education and awareness raising of utilities to the potential for RE systems to contribute to rural electrification and the potential for net metering	●	●	○	○	○	L
Dissemination of information on market support policies	○	●	○	○		L
Introduction of RE education in schools and specific technical education in universities (internationally recognized qualifications)	●	○	○		○	M
Training of trainers	○	○	●	●	○	L
Training of local people in technical, maintenance and business management skills	○	○	●	●	○	L
Awareness raising of the benefits of RE to professionals in social sectors (health, education, rural development, etc.)	●	●	●	○	○	L
Awareness raising in local communities via appropriate media	●		○	○	○	L
Financing and investment						
Provision of grants and soft loans for feasibility studies, pilot and commercial-scale demonstration projects	●	●	●		○	M
Assessment of existing finance schemes and replication and expansion where successful	○	○	●	○	○	H
Training for financial institutions in RE project appraisal (risk assessment) and micro-financing packages	○	●	●	○	○	L

Table 5.1: continued

Common actions needed	Developing country government	Industrialized country government	IFIs	Private companies	Civil society	Political sensitivity
Encouraging micro-financing schemes	●	●	●	○	●	M
Provision of backing and loan guarantees for users in rural areas and new SMEs	●		●	●	○	M
Encouraging export credit agencies to support RE	○	●		○	○	M
Encouraging life-cycle evaluation of energy systems, incorporating the costs of externalities from conventional fuels and benefits from renewables	●	●	●	●	○	H
'SME incubator' loans to develop SMEs without sufficient track record to secure commercial loans	●	●	●	○	○	M
Analysis and assessment of how green certificate trading can encourage RE development	●	○	○	○	○	M
Smart subsidies (targeted, pro-poor, time-limited and transparent)	●	●	○	○	○	H
Other						
Collaborative RE technology R&D	●	●	○	●	○	L
Setting up joint ventures to manufacture technology locally	●	○	○	●	○	H
Setting up partnerships between developing country actors, between industrialized and developing country actors, and between public and private sectors	●	●	●	●	○	H
Stepping up RE deployment and manufacture worldwide to gain economies of scale and bring down costs	●	●	●	●	○	H
Encouraging entrepreneurs and agents for change to get involved in the renewable energy market	●	○	●	○	○	L

H = high; M = medium; L = low. ● Main action. ○ Supporting action, assistance.

Table 5.1: continued

Common actions needed	Developing country government	Industrial-ized country government	IFIs	Private compa-nies	Civil society	Political sensitivity
Participatory planning	●	○	○	●	●	L
Encouraging green consumerism and energy efficiency	●	●	●	●	○	M

H = high; M = medium; L = low. ● Main action. ○ Supporting action, assistance.

and relationships are important for the development of renewable energy systems. Private-sector investment is crucial for technology transfer, so actions which raise the awareness and confidence of investors are very important. However, no one action can be pinpointed in all cases as the most important. The particular situation (locally and nationally), assessed on a case-by-case basis, will dictate which of these actions is most important and urgent to help overcome the specific barriers faced.

Figure 5.1 shows the interaction and partnerships between government, international finance institutions, the private sector and civil society that are needed to facilitate technology transfer. Again, different actors lead certain actions and participate in a range of others.

Some actions are more politically sensitive than others, so will be more difficult for certain actors to engage in. The sensitivity reflects the amount of investment required, the level of international cooperation needed or the potential impact on other actors. A general estimate of the political sensitivity of actions is indicated in the right-most column of Table 5.1, but this will vary depending on the particular country. From these estimates the actions least sensitive to address are those relating to institutional capacity building, information exchange, education and training. These are essential for raising awareness and creating demand and the local capacity to make informed decisions to install, maintain and operate renewable energy systems and to plan systems effectively. Most actions that involve policies and legislation are relatively politically sensitive, but potentially have a high impact

Figure 5.1: Partnerships needed for renewable energy technology transfer and market development

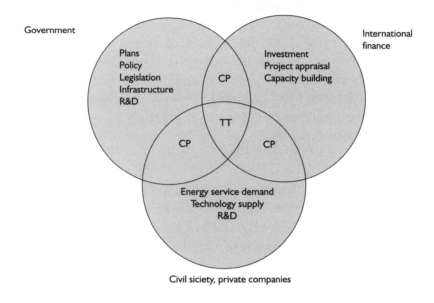

Government

International finance

Plans
Policy
Legislation
Infrastructure
R&D

CP

Investment
Project appraisal
Capacity building

TT

CP

CP

Energy service demand
Technology supply
R&D

Civil siciety, private companies

Key: CP = cooperation or partnership
TT = technology transfer

on confidence building, market development and attracting investment. Some actions (e.g. the CDM and TRIPS) have very high political sensitivity but nevertheless cannot be ignored.

The key actions for each group of actors are summarized below.

5.2.1   Key actions for government

Developing country governments   Host developing country governments can take a range of actions to encourage the transfer of renewable energy technology to their countries. They can help set up the legal and regulatory structures and incentives to engage private companies and other actors in the market. Activities involving information exchange, education and training also constitute an important role for host governments, as does continued involvement in renewable energy R&D.

They can help develop the institutional capacity for public bodies, developing strong, skilled institutions able to undertake energy planning and appropriate policy development and to initiate appropriate regulatory structures. They can integrate renewable energy into PRSPs and NSSDs and other sustainable development strategies, policies or plans, recognizing the role of renewable energy in sustainable development (health, education, enterprise development, gender equity, rural development, etc.). They can encourage entrepreneurs to engage in the renewable energy market by providing information on the potential market, available technologies and applications, and by providing tax breaks and incentives to set up new businesses.

Industrialized country governments   Governments of industrialized countries can take action by encouraging their industries to transfer technology and set up partnerships and joint ventures with organizations in developing countries. They can provide an example to developing countries by removing subsidies on fossil fuels in their own countries. They can take action in almost all of the financial and investment activities as they can influence and back IFIs in their actions. They can assist the export of technology by encouraging their ECAs to extend loan guarantees and export credits to renewable energy technology projects. This can be done by training ECA staff in project appraisal methodologies for renewable energy systems and capabilities to assess the risks (due diligence) and benefits of renewables. Industrialized country governments can also encourage multilateral and bilateral finance institutions to bring energy (and particularly renewable energy) into the mainstream of all projects and to view energy as an integral part of the development process for poverty reduction. They can take action to help in collecting, analysing and disseminating information on technology development, and on fiscal and regulatory policies and their impact on renewables projects. They can take action to influence the development of new incentive schemes such as green certificate trading and green funds, and help negotiate international agreements such as TRIPS and mechanisms such as the CDM so that they give maximum developmental benefits to developing countries and incentives to industry to engage in market development. They can also act as important partners in renewable energy R&D and in establishing partnerships between the various developing country and industrialized country actors.

5.2.2 Key actions for IFIs

International finance institutions can take action to provide access to finance via grants, soft loans and commercial loans (particularly where commercial banks perceive the risk to be too high, but where the projects have a developmental benefit). Indeed, like industrialized country governments, they can be vital actors in almost all financing and investment activities relating to energy efficiency and renewable energy. Finance can be provided for both investment in technology transfer and local capacity building. Their actions are instrumental in many cases for setting up partnerships, as grants and soft loans often make up only part of the full investment needed, so other actors are sought for investment partnerships. They can encourage partnerships between developing country actors as well as between industrialized and developing countries. They can invest in programmes that encourage entrepreneurs to engage in the market and assist the development of SMEs. International finance institutions can support many other actions including the restructuring and privatization of the electricity supply industry in developing countries, the removal of subsidies on fossil fuels and training, and the analysis and dissemination of information. International finance institutions can encourage bundling of small renewable energy projects to reduce transaction costs and to make them bankable by putting lots of projects together in one package for financing.

5.2.3 Key actions for private companies

International private companies can take action to provide technology through retail sales, partnerships and joint ventures, to engage in technology R&D and to train local staff. They can also help set and enforce technology standards and expand technology manufacturing levels worldwide to bring down costs. Local private companies can take action to install, maintain and operate the technologies. Actions of both international and local private companies are important to support most of the information exchange, education and training activities, and in performing market and resource assessments and feasibility studies. In the latter, they can have significant influence on renewable energy uptake through encouraging and undertaking life-cycle costing of energy systems.

5.2.4 Key actions for civil society

Civil society can put pressure on national and local governments to act in its best interest. They can assist in many of the actions including awareness raising, training and dissemination of information. Small local NGOs working at the grassroots level can act to provide micro-financing for remote rural areas. They can also help plan schemes and select appropriate systems using participatory planning methods to empower communities. It is important that representative civil society groups make their voice heard at national level.

5.3 Conclusion

Energy is crucial for economic growth and social development, and demand for energy is rising. There are niche markets where renewables are already the most convenient and economic option for providing required energy services, particularly in remote rural areas where it is too costly to extend the grid and difficult to transport fossil fuels. Renewable energy has a key role to play in meeting international development targets, reducing local environmental impacts, meeting global carbon dioxide reduction targets and ensuring security of energy supply. If the transfer process is to be sustainable, it is important to transfer both the technology and the skills and knowledge to select, adapt, install, operate and maintain the technology.

It is important to attract investment into the renewable energy sector. This can be done by increasing the awareness and understanding of renewable energy systems, by using suitable project appraisal methodologies, by valuing both the externalities and benefits of energy systems, by understanding the risks, and by increasing confidence in renewable energy technology and projects. There is a range of actors involved in financing renewables and the private sector is increasingly playing an important role. However, it must be recognized that there are areas where renewables are unlikely to be attractive to the private sector in the short to medium term, and in these areas targeted 'smart' subsidies will be required.

Attracting investment is only one-half of the issue: finance then needs to be made accessible to those who need it, through a combination of finance mechanisms including micro-financing. It is important to assess the effectiveness of existing finance mechanisms and disseminate best practice. It is also important to continue to develop and support innovative finance mechanisms that encourage the transfer of technology to developing countries in particular, and SME incubators to help SMEs grow and build up a sound track record, enabling them to attract commercial loans.

A potentially large route for investment in renewable energy systems in developing countries is the CDM. With the background of concern over climate change and the need to reduce GHG emissions, renewable energy is well positioned to help achieve emissions reduction targets. The creation of the CDM as one of the Kyoto mechanisms opens up the prospect of investment that could speed up the transfer of renewable energy technology from industrialized to developing countries in return for emissions credits. Even though there are mixed views over the merits of the CDM, potential projects are already being identified for future investment. CDM projects must fit in with the sustainable development goals of the recipient country. Renewable energy projects have good potential to contribute to sustainable development, particularly in rural areas of developing countries, and to help achieve international development targets. The number and size of CDM projects in developing countries will be influenced greatly by the criteria used to select projects and by how the CDM is set up and operated.

There is a range of barriers to renewable energy technology transfer, but they are surmountable, especially if relevant actors are aware of the options for overcoming them. This book has identified common actions that need to be taken to overcome barriers to renewables. They need the involvement of a range of different actors taking either leading or supportive roles. Partnerships can help spread the investment costs and risks, facilitate the exchange of information between partners and speed up the transfer process. Lessons from other countries and projects need to be disseminated and successful projects replicated.

With an increased awareness of the developmental benefits of renewable energy and the contribution it can make to environmental and security issues, investors are becoming more interested and renewables are getting on the political agenda. Successful technology transfer requires attention to commercial management, market development, market competitiveness and technology adaptation. Actions need to be chosen to help achieve one or more of the following aims: bringing down the costs of renewables and making them affordable; increasing confidence in renewables; attracting investment into renewables; developing local institutional strength and building local capacity; developing supporting infrastructure (e.g. SMEs, legal and regulatory systems, transport and communications); and increasing awareness to stimulate demand and market development.

There is a bright future for renewables as contributors to global environmental protection and security of energy supply, as a means of avoiding lock-in to outdated and environmentally damaging fossil-fuel technologies; but the greatest potential they have is to contribute towards sustainable development in developing countries. A wide range of actors from both industrialized countries and developing countries need to work together to realize the full potential for renewable energy.

Annex I

Case Studies

A.1 Introduction to case studies

Two main renewable energy technologies were chosen as a focus through which to look at options for overcoming barriers to the use of such technologies: off-grid solar home systems (SHS) in rural areas; and the export of electricity to the grid from biomass-powered systems (including cogeneration systems producing both heat and power). Two technologies were chosen for ease of comparing lessons learned between different countries. These technologies are not necessarily the most representative technologies for all developing countries, nor do they have the greatest energy generating potential; they were chosen because they cover a broad range of potential barriers and issues. Although both technologies produce electricity, it is recognized that heat, light and mechanical power produced by non-electric technologies are also very important. Many of the non-technical barriers facing electric technologies and non-electric technologies are similar. The two technologies chosen highlight many of the general barriers faced by renewable energy technologies both on-grid and off-grid, but obviously do not cover the specific issues faced by every other technology.

The case study analysis is base on research undertaken for NEDO in Southeast Asia; however, a few other countries have been included where the author had experience and access to information. Case studies have been taken from the following developing countries: India, Indonesia, Kenya, Philippines, Thailand, and Vietnam. Material for this book has been collected using a combination of desktop research and visits to developing countries. When in developing countries, meetings were held with key actors in the renewable energy field, including where possible: policy makers, financiers, developers, technology manufacturers, installers, users, utilities, consultants and NGOs.

A.2   Solar home systems case studies

Case study 1: solar home systems in Indonesia[1]

Project outline   The government of Indonesia (GOI) considers rural electrification to be a means of promoting social and economic development in rural areas. Indonesia's rural population is spread over about 6,000 of its 16,000 islands, so it is not economic to provide all areas with access to primary grid electricity. Thus isolated local mini-grids or individual stand-alone systems can play an important role in rural electrification. SHS have the advantage of being modular, so the size of the system can be increased as demand rises. The systems have low environmental impact at point of use,[2] and are free from fuel requirements as the sun is freely available.

The presidentially assisted rural electrification programme (BANPRES) in Indonesia has installed more than 3,300 SHS in 13 provinces since it started in 1991. This pilot project was funded by a presidential grant through the development budget totalling Rp3.4 billion (approximately US$1.6 million at the time) and employed a revolving fund mechanism. The participating government agencies absorbed the overhead costs. Lessons learned from the BANPRES project have paved the way for commercially orientated initiatives and further government-sponsored programmes. There are now over 100,000 PV systems installed in Indonesia. In 1992 the GOI announced a 50 MWp PV programme, one of the most ambitious PV programmes in the world. This ultimately aims to install around 1 million PV powered supplies in rural households in Indonesia. The overall potential market for SHS in Indonesia has been estimated at somewhere between 2 million and 10 million units.

The BANPRES project was designed to make use of existing rural structures and capabilities. It required collaboration between GOI min-

---

[1] This case study is based on information collected from various sources including ETSU (1999b).

[2] To keep the environmental impacts of solar home systems to a minimum, it is important to set up a battery recycling scheme and encourage people to return old batteries to recycling points. This can be done by having a refundable deposit on the batteries. Even if the owner of the old battery is not interested in returning it for recycling, it is likely that someone else will return it, taking the opportunity to receive the deposit. If batteries are recycled the disposal of electrolyte (sulphuric acid) can be dealt with properly and the old lead can be mixed with new lead to produce new battery plates.

istries, local government and the recipient villages to ensure technical, socio-economic and financial objectives were met. However, due to the number of players in the project, some communications problems were encountered, and the decision-making process was slow and bureaucratic.

The project leader for BANPRES was the non-departmental government agency for application and assessment of technology (BPPT), which was responsible for technical support including defining the technical specifications for the SHS. In association with the Ministry of Cooperatives (MOC), BPPT established the criteria by which villages were selected for participation in the programme. Factors considered were proximity to grid services at the time and in the future, the community's desire for electricity, householders' ability to pay and the effectiveness of local cooperatives, called *Koperasi Unit Desa* (KUD). The effectiveness of the KUDs was important as they played a major role in implementing and collecting down payments and monthly instalments. Each KUD employed two technicians to provide maintenance services for the SHS. A widespread presence in rural areas, the Bank Rakyat Indonesia (BRI) was part of the institutional organization. BRI collected money from the KUDs and passed it on to the revolving fund controlled by BPPT. Private companies installed the systems and had responsibility for training technicians and providing after-sales support in the form of warranties, spare parts and advanced technical assistance as needed by KUDs. Users were responsible for the elementary maintenance, payment of fees and load management. Long delays were encountered in the repair of systems due to a lack of spare parts. This could have been alleviated if the KUDs had stocked spare parts.

The users signed a lease-purchase agreement with the KUD. The terms of repayment were a down payment of Rp50,000 (approximately US$24) and a monthly payment of Rp7,500 (approximately US$3.6) for 10 years at 0% interest. There was poor enforcement of system removal on non-payment, which led to many more households ceasing to make payments as they saw their neighbours getting away with not paying. The collection rate hovered around 60%. The cost of component replacement (e.g. battery, lights) was the responsibility of the user. Under the terms of the agreement, the KUD retained ownership of the SHS until the loan was fully repaid. Some of the early pilot projects were run at no cost to the users, often through bilateral grants with added government financial support. Sometimes collection of very low 'service fees' was instituted, but the actual collection of such fees in the field was below expectations. This led to the false expectation that such systems could be obtained free or for very low monthly payment rates.

The BANPRES scheme included pilot demonstration and optimization of SHS configurations that have been replicated, virtually unchanged, in both commercial and government-sponsored programmes. Typical systems included a 45Wp or 48Wp solar panel, an automotive battery, a controller, two fluorescent lights and a 12V DC socket. The systems can provide lighting for seven to eight hours, or one light and television for about five hours. Overall the SHS are reported to have performed well.

Lessons learned   The main lessons learned during the BANPRES programme and subsequent SHS projects, which are now being addressed in the further projects that make up the 1 million PV programme, were:

- There was a lack of standardization in the approach taken by the KUDs, resulting in many inconsistencies in book-keeping and collection practices. This led to users paying different amounts. When people realized others were paying less, they also wanted to pay less and in some cases did not want to pay at all.
- It is important to have a strong system removal policy on non-payment to encourage users to keep up payments.
- It is important not to raise expectations that SHS can be obtained free or for very low monthly payment rates, as in reality it is not possible to extend such favourable conditions to all potential users.
- KUDs should stock spare parts, to avoid long delays awaiting repairs.
- When choosing areas to include in SHS projects, it is important to know if and when the state electricity utility intends to extend the grid to that area. The state electricity corporation in Indonesia has been accused of changing its plans for grid extension, and in some cases electrifying solar villages within two years of the SHS being installed, instead of the previously planned period of five to ten years. Households then prefer to use grid electricity and are unwilling to continue paying for their SHS, leaving loans unpaid.
- The potential market estimates for SHS and the government-backed 1 million PV system programme are powerful incentives for SHS dealers to get involved in the market – particularly as several large

bilateral and multilateral SHS projects have been announced since 1992, for example by AUSAID (36,400 SHS), the World Bank/GEF (200,000 SHS – though this was cancelled)[3], BIG-SOL (Bavarian bilateral aid: 30,000 SHS and 300 central PV systems) and DINO-SOL (Dutch bilateral aid: 35,000 SHS; this too was deferred, but the Sol-Invictus programme is continuing)[4].

- Lack of credit schemes has been a major barrier to the market potential for SHS in Indonesia. It should be noted that users are not the only actors that might require access to credit; SHS dealers, installers and other key actors may also need credit. Also, the conditions for credit may be difficult to meet. For example, SHS dealers must prove their creditworthiness and performance history in order to secure a loan. This is difficult if they are new organizations trying to enter the market. Banks therefore need to address the criteria by which they assess creditworthiness for different actors e.g. dealers, installers, technology manufacturers and users.

- SHS dealers found the BANPRES project very bureaucratic due to the complexity of working with cooperatives. This resulted in dealers having little control over what they did and how they did it, making them reluctant to be involved in similar projects. They prefer to deal directly with customers and avoid community schemes and government programmes, which they see as problematic.

- SHS dealers found it difficult to use agents in the field to collect payments for private contracts, as they are not easy to control or audit. One SHS dealer spoken to preferred to set up his own network of service centres in the field, but this can only be done if there is a large enough market for SHS in the area (500 SHS or 5% of the households). The dealer trained three local staff for each centre, one for marketing, one for installation and maintenance and one for credit programmes.

- Lack of SHS supply and service chains in rural areas means users cannot benefit from economies of scale and competition among SHS dealers, which would bring down prices. The lack of supply

---

[3] The World Bank 200,000 solar home system project in Indonesia was cancelled in early 2000, mainly because of problems created by the economic crisis in Asia.

[4] The DINO-SOL project was deferred in 1999 mainly due to problems created by the economic crisis in Asia; however, the Sol-Invictus programme, whereby the Netherlands government is providing an export subsidy for 20,000 PV modules for Indonesian solar home system installations, is continuing.

and service chains means the dissemination and expansion of the market from urban to rural areas does not take place readily. Also, when systems need replacement parts, it takes a long time to obtain them, possibly resulting in the system falling out of use.

- Users were not given adequate training in basic operation, maintenance and understanding of their SHS. In some cases users disconnected their controllers to get more power from their batteries. Neighbours, seeing what they had done, followed suit; unfortunately, such abuse leaves the battery permanently damaged after a short space of time. Users also attempted to charge a second battery for themselves or a neighbour. Depending on how and when the second battery is connected, it can result in either the first battery not getting fully charged (which reduces the life of the battery in the long term if it continues) or damage to the controller (if they connected the second battery load side of the controller, as it will attempt to draw a large voltage, overheating the controller). The controller is set to deliver 12V DC load side, and a 12V DC battery needs a voltage above this to charge successfully.
- It is important that households have a sense of ownership of or value in the SHS to motivate them to use and maintain them carefully. In projects where households paid little or nothing for the systems, misuse or damage were more common than in projects where households had paid the full cost of the systems.
- Households in remote areas have difficulties with cash payments. It would be easier for them to pay in commodities e.g. crops, livestock, handicrafts etc. It is also difficult for them to pay regularly each month as this does not coincide with incomes from harvest etc.

Case study 2: solar home systems in the Philippines[5]

Project outline PV technology has until recently been seen in the Philippines primarily as a pre-electrification technology rather than a permanent energy solution. In 1987 the Special Energy Programme (SEP) was initiated under a bilateral agreement between the German government and the government of the Philippines. The strategy of this programme was to install SHS in clusters, gradually increasing the

---

[5] This case study is based on information collected from various sources including Cabraal et al. (1996).

demand for electricity and the area covered until grid extension became an economic alternative. However, the population of the Philippines is spread over approximately 2,800 of its 7,100 islands and islets, and consequently mini-grids or stand-alone systems are the more practical and economic route to electrification in many cases.

The SEP project included funding from the National Electrification Administration (NEA) to the rural electricity cooperatives. Revolving funds were set up to enable further systems to be financed. The rural electricity cooperatives were responsible for quality, servicing and fee collection. Local technicians from the cooperative were generally responsible for installation, troubleshooting and maintenance. Where and when access was difficult for the local cooperative, a local NGO was identified to assume responsibility for carrying out the servicing, monitoring, safety and fee-collection elements of the SEP project.

SHS were bought in quantity by the SEP on behalf of the local rural electricity cooperatives, thus obtaining bulk discount prices for the equipment. The cooperatives sold the systems to the users, who had to pay cash for the balance of the system (excluding the controller and PV panel). The users then made monthly payments to cover the cost of the controller and PV panel. The SHS components were exempt from import duty and value added tax (VAT). Battery recycling in SEP was effective as users had to bring along their old battery in order to purchase a new one from the rural electricity cooperative.

A typical system included a 53 Wp PV panel, a deep discharge lead-acid battery, a controller, a DC/DC voltage converter and five lamps. The systems on average would produce enough energy to power one fluorescent light for three hours, one compact fluorescent light for four hours and one radio for 12 hours. Locally manufactured components such as batteries, controllers and fluorescent lamps performed poorly, but this was attributable to poor quality of manufacture or improper use, rather than to the systems being unsuitable for the required services.

The Philippine government is currently developing a framework programme on Energy Resources for the Alleviation of Poverty (ERAP). The programme aims to improve the quality of life in rural areas by providing adequate and sustainable energy services. It has identified the need for accurate renewable energy resource assessments. A wind assessment for the Philippines has been completed, but solar, small hydro and biomass resources still need assessing. It has also pointed to the need to encourage private investment in sustainable technologies.

The Development Bank of the Philippines (DBP) has received specific training (funded under the FINESSE initiative)[6] on project evaluation and appraisal for renewable energy systems. This shows that it considers renewable projects to be important for the future energy requirements of the Philippines and recognizes that standard energy project appraisal methods are not suitable for renewable energy projects.

In combination with the Malampaya deep water natural gas exploration project, Shell has been planning a rural energy service company (RESCO) in Palawan which would provide electricity to rural areas via sustainable energy sources including solar PV. One stumbling block, still to be overcome at the time of writing, is the franchise to be set up enabling the RESCO to sell electricity to users on Palawan. Previously the Philippine government gave sole rights to regional electricity generating companies to supply electricity on each island at a set price equal to that of the grid on the mainland. As grid electricity is subsidized, the price is too low for renewable energy systems to be economic. This has been a deterrent to provision of electricity in the islands. The fixed price may be removed, which will encourage PV and other renewable energy systems as RESCOs will thereby be able to see a return on their investment.

Lessons learned   The following lessons, learned during the SEP project, have proved useful in planning further SHS projects and programmes; those learned during the planning stages of the Shell solar project are also valuable.

- Users of the systems supplied under the SEP received training and a training manual in the style of a comic book that proved to be very effective.
- Regulation regarding the supply and pricing of electricity to users in rural areas can be restrictive and prevent private investors from entering the market.
- Battery recycling in SEP was effective as users had to bring along their old batteries in order to purchase new ones from the rural electricity cooperatives. This was encouraged by placing a refundable deposit on batteries.
- Locally manufactured components performed poorly due to poor quality of manufacture or improper use. The quality and expected

---

[6] See Chapter 3 for more detail on FINESSE.

life of local technology need to be tested and weighed up against the cost saving, to see if in the long run it will cost more as a result of more frequent replacements being necessary.

- Project evaluations for renewable energy projects need to be carried out in a different way from those for conventional power projects. Factors that need to be taken into account when evaluating renewables projects include the relatively high up-front cost and the difficulty in establishing long-term biomass fuel contracts. Environmental benefits compared to fossil fuels should also be taken into consideration.
- A clear government policy on renewable energy is important to set the foundations for renewable energy development in developing countries.
- It is also important to have clearly defined roles for government departments that are involved in renewable energy, preferably with one department identified as the 'champion' for renewable energy.
- It is important to recognize that SHS have a role to play both in developing a market for electricity via pre-electrification institutions, and for stand-alone applications on a long-term basis (e.g. SHS or battery-charging stations).

Case study 3: solar home systems in Vietnam[7]

Project outline   In 1994–5 the first phase of the Vietnam Women's Union (VWU) PV project was set up, supported by the Solar Electric Light Fund (SELF), with technical assistance from the Institute of Energy (IOE) in the north of Vietnam and Solarlab in the south. This was a three-year pilot project in two provinces and within four communes. The main aim of the project was to improve conditions for women in rural areas, but such improvement would of course have benefits for the rest of the family.

Training courses were held for local VWU commune staff, and manuals were produced on the operation and maintenance of the systems. In this pilot phase all the equipment was provided by SELF, which also provided funds for project management and collection of fees. Users had to pay 20% of the system cost as a down payment, then the remainder by monthly instalments over three years. For example, a 40 Wp

---

[7] This case study is based on information from various sources including ETSU (1999b), Wilkins et al. (1998) and personal communication with Neville Williams (SELCO).

system cost approximately 4.5 million dong (approximately US$500); therefore the down payment would be 0.9 million dong (approximately US$100) and monthly payments 0.1 million dong (approximately US$11). The VWU collected payments for the systems and put the money into a fund ready to finance management of a second phase of the project.

By the end of the three years, 350 PV systems had been installed (including five community systems of 220 Wp and SHS ranging from 22.5 to 40 Wp). The SHS provided sufficient power for lighting and electricity for a radio, cassette player or television, thus improving the quality of life by providing entertainment, access to information and good-quality light for reading and household activities (cooking, eating, looking after children, sewing, handicrafts, socializing).

There is a good track record for collection of payments. By late 1999 over 80% of households had completed their three years of payments and therefore owned their systems. The remaining 20% have delayed repayments because their systems have broken down or because they are not able to make payments regularly. This pilot phase proved that payments could be collected and the electricity requirements of people in rural areas could be met at an affordable price.

Locally produced batteries were found to be of poor quality, failing after only six months in some cases (they should last for around five years). This might, however, have occurred because distilled water was not used to top up the batteries. For replacements, imported automotive batteries from the United States were used, and these have lasted for four or five years so far. Deep discharge batteries purpose-made for SHS are designed to have a life of around eight years.

The second phase of the project started in 1998 with a plan to install 500 PV systems in rural areas. This phase was more difficult as there were no subsidies for the systems, so loans had to be arranged with a bank. Prior to the second phase, awareness of potential users had to be raised; however, the high initial cost of the systems was seen as a barrier and many households were unwilling to take out a loan.

The developer in the second phase is SELCO-Vietnam, a commercial solar service provider owned by the Solar Electric Light Company (SELCO) of Washington DC. The Vietnam Bank for Agriculture and Rural Development (VBARD) provides loans to users. The VWU collects payments on the loans on behalf of the VBARD. Local government undertakes the credit screening, confirming that households are creditworthy, reliable and live at the address intended for the SHS to be installed.

System contracts are set up between the VWU, the user and SELCO-Vietnam. The user contracts were stricter in the second phase than in the pilot phase, and the threat of action to remove systems after two to three months of non-payment has proved effective, causing payments to be made more regularly.

SELCO-Vietnam secured an IFC loan of US$750,000 for seven years at 2.5% interest to provide consumer credit. This will enable SELCO-Vietnam to pay US$50 into VBARD for each system it installs as a guarantee against which the user can take out a loan for 75% of the system cost. VBARD normally provides loans only to projects that show an increase in agricultural production, or a direct income-generating benefit, but is willing to provide loans for the SHS given the VWU's involvement in the project, SELCO-Vietnam's loan guarantee and the local government credit assessment. If the PV module suppliers were to guarantee that the modules have a value depreciating over time, for example by offering to buy back modules at an appropriate depreciated value if households defaulted on their loan payments, then the value of the modules could act as a loan guarantee and SELCO-Vietnam would not need to find funding for this separately.

Households have to make a 25% down payment and take out a loan for the remaining 75% at 1.15% interest per month. Vietnam introduced VAT at 10% on all goods in January 1999, but SELCO-Vietnam's SHS were exempted from the tax in 2001. SELCO-Vietnam, together with the VWU, had installed nearly 2,000 SHS by the middle of 2001.

Lessons learned

- There is a lack of a comprehensive government policy on renewable energy. Clear policies on the development of renewable energy act as a signal to private organizations that there is a market for renewable energy systems. The government is aware of the importance of renewable energy in rural development and has created some policies to help; for example, work is being done on setting standards for PV systems and developing skills and technology to manufacture PV system components in Vietnam.
- The backing of a well-respected organization (the VWU) and local government involvement are important for the credibility of the project and in securing loans for users.

- The role of the developer, in this case SELCO-Vietnam, is key to the project getting off the ground. VBARD was unlikely to give loans to users without the guarantees offered by SELCO-Vietnam, and the negotiations between SELCO-Vietnam and VBARD to get the bank involved took a long time. SELCO-Vietnam is now thinking of setting up a 'green fund' to give loans to users, avoiding the necessity of going through the banks and thus speeding up the process.
- The involvement of the VWU in collecting fees is effective as social pressure is put on the women of those households that do not pay. For example, women are not allowed to attend VWU meetings if they are behind in their payments.

Case study 4: solar home systems in the South Pacific[8]

Project outline   Kiribati and Tuvalu are independent island states in the South Pacific. Kiribati is made up of 33 islands of which 17 are inhabited. The total population is around 80,000. Tuvalu consists of a group of nine coral atolls, all of which are inhabited, with a total population of around 10,000. In late 1991 the PV follow-up programme to the CEC-funded Pacific Regional Energy Programme (PREP) was designed for five countries in the South Pacific (Kiribati, Tuvalu, Fiji, Tonga and Papua New Guinea). Under this programme 250 SHS were installed in Kiribati, on the islands of Marakei, Nonouti and North Tarawa, between 1993 and 1995. Installations on the outer islands of Tuvalu included 50 SHS, 7 larger (pilot) PV systems with domestic refrigerators and 8 vaccine refrigerators, along with the upgrading of 226 existing SHS.

An international call for tenders to supply the necessary equipment was launched in 1992. Equipment was ordered in quantity for the whole programme by the Forum secretariat overseeing the programme in Fiji. This allowed bulk discounts to be obtained. Equipment was delivered during 1993 and 1994 and installed by the end of 1995 (Gillett and Wilkins, 1999).

An important part of the PV follow-up programme was a series of workshops and training sessions for solar companies and cooperatives in the participating countries on the preparation of tender documents, tendering procedures, the evaluation of tenders, and technical skills.

---

[8] This case study is base on information from various sources including Gillett and Wilkins (1999).

These provided significant institutional strengthening and helped the Solar Energy Company (SEC) in Kiribati to develop the skills necessary for the local production of electronic controllers and DC/DC converters. They also provided the opportunity for people involved in the project from each country to meet and exchange experiences and ideas.

❑ Kiribati: The SEC (a financially independent government-owned organization) was responsible for inspecting, testing and provisionally accepting the equipment when it arrived in Kiribati. The SEC was contracted by the Forum secretariat in Fiji to install the systems and train the staff and users, with technical assistance provided by the project consultant, the South Pacific Institute for Renewable Energy (SPIRE). A full inspection of all systems by SEC and the project consultant was carried out after installation. This was important, ensuring that any installation mistakes or problems were identified early and corrected. Also it was a very useful learning process for the local technicians and engineers.

The SEC trained a technician for each of the islands on which SHS were installed. The technician is responsible for regular maintenance of the SHS and for collecting the monthly payments. The SEC has an engineer on the main island who can be contacted if the technicians come across any problems, and who visits the other islands periodically.

Users had to pay a deposit of A$50 (approximately US$29) to show their willingness to pay for the PV systems, and then monthly payments of A$9 (approximately US$5) per month. The systems remain in the ownership of SEC and the monthly payments go towards maintenance costs and replacement of equipment. The users have to buy replacement light bulbs, but the SEC replaces wiring, controllers, DC/DC converters, switches, light ballasts and batteries when needed. If users fail to make payments for three months, they are to be disconnected and their systems are removed. Their debt is taken out of their deposit; if they wish to get a system installed again, they have to go to the bottom of the waiting list and also pay the deposit again. There are not enough systems to meet the demand, so the waiting list is long.

A typical system consists of two 50 Wp PV panels, a 100 Ah deep discharge battery, a controller, three lights, three switches, one night light and wiring. A DC/DC converter for a radio cassette deck is optional for A$1 (approximately US$0.6) extra per month. Systems were still working well five years after installation. The project has provided benefits for the users as intended (lighting in the evenings for reading, handicrafts, tending to children and the sick, etc.). The project was very well managed, with dedicated staff at the SEC and an expert providing technical assistance all determined to make the project successful.

❑ Tuvalu: The Tuvalu Solar Energy Cooperative Society (TSECS) installed similar equipment around the same time, but the systems have a higher failure rate which can be attributed mainly to poor management. The cooperative structure and management worked well in the beginning, but unfortunately corrupt management stole funds from the cooperative society, leaving it without sufficient money to replace equipment when it broke.

The monthly fee is lower than in Kiribati at A$7 (approximately US$4), but the deposit is the same. In 1999 the manager of TSECS tried to raise the monthly fee by a small amount to a more realistic rate that would cover the costs of the TSECS staff who manage and maintain the systems (though it would not have made up for the stolen capital, which should be there to replace equipment). However, the power to decide on the monthly rate lies with cooperative representatives from each island, and they refused to increase it.

It is very interesting to see how much people value their systems. On the island of Nukufetau in Tuvalu there is a mini-grid based on diesel generation. This runs for a few hours in the morning and a few hours in the evening as it is too expensive to operate for longer. While it is switched on, people watch TV and video, or chill down their freezer, but not all houses are linked up to the mini-grid. Some houses have no electricity, some have PV or grid, and some have both. Those with both value having the solar electricity as it is available any time of day or night, rather than being restricted to certain hours of operation. This can be particularly useful when dealing with sick children at night. Those households with just PV either cannot afford both types of supply, or choose not to have grid electricity. One reason given for not wanting grid electricity was that the owners were concerned about the safety of their grandchildren (grid electricity is of higher voltage and several children were known to have electrocuted themselves).

Lessons learned
• It is very important to remove systems promptly on non-payment. Often people choose not to pay rather than not being able to pay.

Because the technician lives on the island, in the small community, he will be aware if his customers are suffering real financial hardships or family problems hindering payment, and can take appropriate action – such as rescheduling payments if they are really unable to pay, or cutting them off if they are able to pay but are not paying. Many families have sons who are employed as seamen; they send money home to their families, but at irregular intervals. One elderly couple in Kiribati paid a whole year's fees in advance, as their son had given them the money to do so; they were afraid of spending the money on other things, and valued their SHS enough to pay upfront to secure its use for the year.

- The person selected to be the technician on each island must be chosen carefully. If too young he will not command enough respect within the community to collect monthly payments from the households. Whoever the technician is, he will come under social pressure, as in small remote communities he will inevitably be related to a large proportion of the households with PV systems. This makes it difficult for him to enforce system disconnection or removal, and to insist on payments being made. Clear contractual conditions for households should be stated, making it obvious what they can and cannot do and what the consequences of their actions will be (e.g. they must pay regular monthly instalments or they will be cut off; they must not tamper with the system or it will be removed). Involvement and back-up from SEC headquarters on monitoring and technical support can help to neutralize local social pressures on the technician.
- In Kiribati, the technician carries out regular maintenance on the system each month when he collects payments. This keeps the batteries and systems in good working order and enables him to detect any problems early, preventing permanent damage to components. He is encouraged to keep the systems in good working order as households do not have to pay when the systems are not working and his salary depends on the fees he collects.
- The training and capacity building provided as part of the programme were very important as they ensured that the tendering and selection of equipment were carried out in a professional manner, and that equipment standards were drawn up and met.
- Coordination of the programme throughout the region allowed bulk purchase of equipment and reduced prices. However, tied aid was

used to fund the project and this restricted the sources of equipment and the choice of contracting companies; therefore, the most economic technology may not have been accessible. This may make maintenance and spare parts more expensive in the future.

- Technicians have only basic equipment and often forget to carry their screwdriver or other tools with them, or lose them. It is therefore important that the batteries are easy to open, preferably of a kind on which you can unscrew the cell lids by hand rather than needing a screwdriver.

- Batteries are often tucked away in dark corners of the house. See-through battery sides make it much easier for the technician to verify water levels in individual cells.

- Rain water was used to top up the batteries as it was tested and proved pure enough. This removed the need to transport bottled water to the island, or to install distillation equipment. It must be noted, however, that the use of rain water is not likely to be suitable in many locations as it depends on the collection method used: the water can easily be contaminated with dust or other things with which it comes into contact.

- The SEC controllers were designed for easy maintenance and to withstand the harsh corrosive marine environment on the islands. Other high-tech controllers were tried in the field, but were found to corrode, and when they needed repair it was impossible for the technicians to do so with the basic equipment to hand. The high-tech systems were too small and intricate. SEC controllers were exported for use in Tuvalu and other Pacific island states as well.

- The monthly fees collected in Kiribati are used to cover SEC staff and overhead costs; the remaining amounts are invested to gain interest. Other activities, such as controller production, are needed in order to top up the income for investment so that equipment can be replaced when its useful life is up. The number of systems in Kiribati needs to expand to at least 1,000 in order fully to cover the above costs without recourse to other sources of income. The European Commission has agreed to fund another phase of instalments to increase the number of systems to this level.

- There is no battery recycling scheme at present. This is important to protect the environment on the islands. There is very little land and no room for garbage disposal. Ground water is the only source of fresh water in Kiribati, and it must be protected from contamination.

It would be good to include other small batteries used in torches and radios in any recycling programme, and perhaps plastics and tins as well. One big constraint on recycling is the frequency of transport to and from the islands, and the unwillingness of boat owners to take the batteries on board, as they contain acid and the owners are afraid of any spillage causing contamination of food cargo or corrosion of the boat.

Case study 5: solar home systems in Kenya[9]

Project outline   Kenya is one of only a few countries in the world which can claim to have a self-sustaining commercial PV market that is essentially free from government intervention. The solar PV sector in Kenya started to develop without the help of government programmes or large aid-funded projects. The PV industry is small but dynamic, growing between 1982 and 1999 to a value of US$6 million per year. Over 120,000 SHS have already been installed and the market has continued to grow at around 12–18% per year since 1992. Government involvement in the development of the PV sector in Kenya has for the most part been restricted to specific projects, for example PV power supplies for schools and rural clinics, for telecom relay stations in remote locations, for rural water supply systems, for high-frequency radios and for railway signalling equipment. To date, the government has not been involved in rural electrification projects involving PV systems.

The PV market in Kenya developed along purely commercial lines in three phases. First, both upper-middle-class rural 'innovators' and off-grid NGO projects installed PV systems which in turn created additional demand for the technology. Second, large numbers of rural people bought small PV panels and batteries primarily to power lights and televisions. Third, hire-purchase and finance agencies began to offer systems, allowing a greater number of rural people to purchase them on credit.

Over 70% of the population in Kenya live in rural areas. There are over 3 million homes outside the reach of primary grid electricity and increases in grid capacity are struggling to keeping up with expansion in demand in urban areas. A large proportion of rural households want

---

[9] This case study is based on information from various sources including ETSU (1999b), and personal communications with Peerke de Bakker, Kenya.

electricity for lighting and amenities such as TV, radio etc. Thus there is a large market for stand-alone energy systems in rural areas. It is reported that Kenyan families typically spend 5–10% of their income on purchasing energy services. It is estimated that about 1 million households can afford to purchase a PV solar home system, and with the cost of PV decreasing the potential market is likely to grow. Although PV systems are still relatively expensive, they can be more cost-effective for household demand or small-scale community services in remote areas than grid extension.

The SHS market was stimulated in 1989 by the importation of low-priced, low-power-rated panels. By 1999 it was estimated that 3% of rural households had installed small-scale SHS for lighting and that at least 70% of the population knew what a PV system was.

One of the characteristics of the PV market in Kenya is that a significant fraction of the systems are bought in small parts with cash, without access to a formal credit and finance system. This approach overcomes the difficulty of generating the initial capital for a system in one go. It allows purchases to be made according to the cash flow of the family, and bypasses the need to establish a credit rating. However, it is reported that around one-third of the systems are bought without a controller, leaving the battery open to abuse. In this way the user trades off a shorter battery life against a cheaper initial cost. Ultimately the user will pay more as the battery will need replacing more frequently. Although this may fit in better with their cash flow, it may end up creating unsatisfied users and damaging confidence in the technology.

The presence of this significant and growing market for SHS has led to the growth of supporting infrastructure. There are approximately 15 main importers and distributors of PV equipment and around 100 agents across the country. System components are widely available and sold through an increasing number of local agents, who account for 50% of module sales and 75% of battery sales. This has resulted in a wide acceptance and awareness of the technology: around 73% of all households are interested in installing SHS or have already done so.

The significant reasons for the strong development of the SHS market in Kenya to date are:

• the presence of a strong cash economy – the initial purchasers tended to be rural people with permanent homes and cash available from the sales of agricultural produce;

- the presence in rural areas of installers and retailers – most operations are centred on Nairobi and Mombasa, but the most successful companies have established local networks for installation and repair;
- the availability of equipment – the strength of the market in Kenya means that constant stocks of equipment can be carried, so spare parts are readily available. The local manufacture of lamps and batteries, which started in the 1980s, has brought down prices; although locally produced technologies are poorer in quality and design, this does make local repair easier;
- high demand for television and radio – rural households aspire to owning televisions and radios. Chinese 12V televisions are widely available for around US$50, and SHS are proving to be a suitable and convenient way to power them.

Equipment imported for conventional electrification projects in Kenya and PV systems purchased for ODA projects are exempt from import duty and VAT. However, commercial PV system components are liable for import duty and VAT at variable rates as there are no clear guidelines as to what taxes should apply. A purchaser may pay as much as 44% tax on a system installed by a distributor. However, if only the PV modules are imported this can fall to 10 to 20% tax.

There is currently a lack of quality standards for PV technology, which is leading to reduced performance and lower than predicted lifetime for systems. There are several reasons for this:

- hardware – some equipment on the market is below internationally accepted standards and other equipment is not appropriate for local conditions;
- local design, installation and maintenance practices have not been standardized;
- there is no systematic training of manpower in the PV field;
- there has been little education about PV in rural and off-grid markets and the situations where it is viable.

The World Bank energy sector reform and power development project in Kenya includes an activity to design standards for PV systems and components. With World Bank funding, the Department of Energy started a study of PV standards in July 2000. Recommendations from the study will be put forward to the Kenyan Bureau of

Standards. Standards must be set carefully, as if they are set too stringently they could stifle the market in the cheap, low-power-rated systems that have hitherto so effectively boosted the PV market in Kenya. Purchasers should be allowed to choose between a large or a small system, and between an expensive, efficient module or a cheaper, less efficient one. Technology needs to be labelled and certified so people know what they are really getting for their money. Warranties, although given for some products, are rarely actually provided in practice. This needs to be tightened up, along with the provision of standards, so that consumers are better protected as well as better aware of what they are buying.

The introduction of 12 Wp amorphous panels in 1989 represented a major stimulus to demand. Though of a lower efficiency than crystalline cells, these thin film modules are available in smaller sizes, enabling a larger group of people to buy systems. This encouraged an increasing number of over-the-counter customer-installed sales. Average SHS size decreased from 40 Wp in 1986 to about 25 Wp in 1996. About 9,000 amorphous modules of between 12 and 22 Wp were sold in 1996, the bulk of which went to one-module systems. Approximately half of the modules sold in 1999 were still 20 Wp or smaller.

The World Bank project has initiated the liberalization of the energy utility in Kenya. The process of liberalization should reveal the true costs of electricity generation and distribution, which are often hidden in large, state-owned utilities. As the true costs of electricity supply are revealed, utilities begin to recognize the economic value of mini-grids and stand-alone systems (such as SHS) for power supply in remote areas.

One of the main reasons why the potential PV market has not yet been reached in Kenya is a lack of sufficient capital to purchase the equipment. If credit systems were available to rural households then a greater proportion of the potential market might be realized. The PV Market Transformation Initiative (PVMTI) sponsored by the International Finance Corporation (IFC) should have a major impact on market development in Kenya. It will inject US$5 million into the market, providing loans and grants to consortia of market players. Early initiatives have concentrated on establishing appropriate credit mechanisms, for example using established financial institutions as a vehicle for making soft loans available to cooperatives, which then purchase approved equipment meeting defined standards.

Lessons learned

- Raising the awareness of households in rural areas is a very important step in developing the market.
- Even though there are some households in rural areas which are relatively cash-rich and can afford SHS, there are still very many households who aspire to have SHS but do not have access to enough cash or credit to buy a system. One of the main reasons why the potential PV market has not been reached in Kenya yet is a lack of sufficient capital to purchase the equipment. If credit systems were available to rural households, then a greater proportion of the potential market might be realized.
- Import duty and VAT can make up a significant percentage of the price of SHS (up to 44%). The government could reduce or remove import duty and VAT on environmentally sound technologies if it wished to encourage their use. On the other hand, levying import duty on systems could be seen as a way to encourage domestic manufacture of components.
- Selling the SHS in affordable units can help the market grow.
- The establishment of an extensive retailer/installer/maintenance network to support the market is crucial, allowing users to access spare parts and maintenance services promptly. It is best to build up the network using existing related businesses in the local areas. When the market is big enough, the PV activities can become a core part of the business.
- Small systems can fulfil an important function and should not be excluded from support activities.
- The desire for access to television and radio is strong and should not be overlooked in favour of lighting-only systems (in 1998 many more systems were sold that were able to provide power for TV, because users wanted to watch the football World Cup).
- System design should consider the trade-off between price and performance. The key factor is the requirement of the user, not the performance the financier desires.
- It is important to have recommended system standards to guide companies and purchasers in their design and choice of technology, but it is not always appropriate to impose those standards, as it could stifle the market. Standards should guarantee performance against a norm; they should not be used to set artificially high norms.

- A certification system is important to ensure that the equipment builds and retains a reputation for reliability. Purchasers tend to make decisions purely on price grounds and are not aware of what differentiates a good system from a bad one.
- It is not likely in the short term that SHS components can be manufactured any more cheaply in Kenya than the components being imported from the Far East. Wages are relatively high in Kenya compared to those in East Asia (due to the Asian economic crisis) and East Asia has established strong design and manufacturing capabilities. However, if it becomes economic to manufacture more components locally, this will create employment and help develop a local skill base. There is some scope for joint ventures or licensing agreements, particularly for PV applications such as PV compressors or PV incubators.[10]
- Local companies can design small-scale SHS, but need to develop more skills in designing larger professional systems (e.g. for health centres, schools, street lighting, small businesses etc.) and also skills in procurement. Procurement is a major challenge and can be more important than system design.

A.3   Biomass cogeneration case studies

Case study 6: rice husks in Thailand[11]

Project outline To meet increasing energy demand, the government of Thailand has an interest in developing electricity exports to the grid from within the agricultural sector. There are additional benefits from using crop residues, including environmental gains from reduced waste streams, displaced use of conventional fuels and a general strengthening of the agricultural sector. In order to help meet its national goals for power generation, Thailand has a continuing programme of power sector reform and liberalization. A key objective of this reform programme is investment in new power generating capacity by the private sector.

To encourage private-sector participation, the government is privatizing subsidiaries of the national utilities, has established an independent power producers (IPP) scheme, and is encouraging small-scale projects

---

[10] Personal communication with Peerke de Bakker, Kenya.
[11] This case study is based on information from various sources including ETSU (1999b).

through a small power producers (SPP) scheme. Both are competitive bid schemes offering incentives and model power purchase agreements (PPAs). IPPs and SPPs receive a capacity payment and an energy payment under the PPA for supplying firm or non-firm capacity. Support is also available via the Energy Conservation Fund (ENCON) programme, which gives grants to sustainable energy projects (e.g. producing biogas from pig manure).

Financial support for energy from biomass demonstration projects has been available from bilateral and multilateral ODA. For example, the EC–ASEAN Cogen programme aimed to accelerate the implementation of proven technologies generating heat and/or power from wood and agro-industrial residues through partnerships between European and ASEAN equipment manufacturers.[12] The programme gave advice and training, conducted pre-investment studies, and undertook seminars and study tours to raise awareness of the technology. Pilot projects supported by the Cogen programme and other organizations are very valuable in showing that the technology can work and building up the confidence of potential developers and financiers. However, foreign-funded demonstration projects are usually tied to foreign technology, which in the long run is not so easy to maintain with spare parts.

The agency responsible for setting the direction of Thai energy policy is the National Energy Policy Committee (NEPC), a cabinet-level committee that determines policies governing fuel and electricity. The operating arm of the NEPC is the National Energy Policy Office (NEPO), which is under the aegis of the Office of the Prime Minister. The NEPC is responsible for overseeing the kingdom's three electricity utilities, the Electricity Generating Authority of Thailand (EGAT), the Metropolitan Electricity Authority (MEA) and the Provincial Electricity Authority (PEA). EGAT produces virtually all of Thailand's electricity and MEA and PEA distribute it.

The Thai government is pursuing policy objectives of deregulation and privatization of the electricity sector in order to

- improve the efficiency of the electricity supply industry;
- reduce the government's investment and debt burden;

---

[12] The EC–ASEAN Cogen programme was run by an organization called COGEN. COGEN stopped receiving EC funds for grants and such activities as mentioned in the text and became part of the Asian Institute of Technology in 2000. At the time of writing COGEN still provided advice, training and technical support for a fee.

- encourage more private-sector participation in the form of IPPs;
- utilize energy resources in the most efficient way.

The government's aim is to have electricity generation, transmission and distribution clearly separated and run by limited companies in the future.

The Department of Energy Development and Promotion (DEDP) is the lead body coordinating the government's renewable energy resources programme. It reports to the Ministry of Science, Technology and Environment (MOSTE), but has little involvement in commercial activities such as SPP.

There is potential for large amounts of excess electricity to be exported to the grid from agro-industries (e.g. sugar mills, palm oil mills, rice mills and sawmills). This has attracted considerable interest from both Thai and foreign developers and from Thai agro-industry, and a number of projects are exporting surplus power to the grid. However, contracts are predominantly for relatively small amounts of non-firm capacity, and the scheme has not created major new investment in biomass-fuelled power generation. Under a non-firm contract, mills have few obligations and can deliver power as and when it suits them. The energy payment is not much different for firm and non-firm capacity (in February 1998: firm Bt1.74/kWh, non-firm Bt1.67/kWh). It is not worth it for the mill owners to contract for a small proportion of firm capacity, because the operations are such that they might find it difficult to meet even limited obligations and thus could face penalties.

Regulations allow SPPs to sell power to EGAT at no more than 90 MW each. As of April 2000 only 180 MW out of 1,580 MW of power generated by SPP was produced from renewable sources, most of it from agricultural residues. In an effort to boost electricity generation from renewables, NEPO announced its intention in June 2000 to purchase 300 MW of power using renewable energy sources from SPPs. It is intended that the ENCON fund will be used to offer successful bidders subsidy contracts with an incentive payment for approximately five years. The incentive payment will be in the form of a pricing subsidy per unit of electricity (kWh) sold to EGAT (or incremental sale in the case of an expansion of an existing project) on top of the SPP purchase prices. The scheme is similar to the Non Fossil Fuel Obligation (NFFO) in the United Kingdom; however, contracts for the NFFO were offered for up to 15 years. Guaranteeing a market for power produced from a scheme over a number of years is of great benefit as it gives more confidence to financial institutions to invest in the projects, especially if the project payback period is less than or equal to the guaranteed market period.

The price paid by EGAT for energy produced from biomass by the SPPs is based on the price it pays for heavy fuel oil (i.e. world market price). This rose dramatically with the devaluation of the baht from August 1997. Before the Asian economic crisis of that year the price of heavy fuel oil was Bt2.8/litre and the corresponding energy price to EGAT was Bt0.87/kWh. By February 1998 the price for oil was around Bt6/litre and the energy price was Bt1.67/kWh.[13] If the increase in cost cannot be passed on to PEA and hence the consumers, then the SPP scheme becomes a lot more expensive for EGAT.

The privatization of PEA has been approved by cabinet in principle, but the actual process of change is slow. SPPs deliver their power to PEA, which is involved in the connection of the SPPs to its system, although connection and metering are mainly EGAT's responsibility. The connection costs are borne by the SPP developers (some of which complain the costs are too high). PEA is looking at opportunities to invest in SPPs, particularly in remote rural locations where providing grid-connected power would mean incurring a large financial loss. In these areas transmission losses can be very great. Stand-alone systems or mini-grids (e.g. mini-hydro, solar PV, biomass power generation, diesel generators) are more economic in these regions than grid extension.

Direct third-party sales between companies are not possible at present. However, if the power generator and consumer are the same company and are located close to each other, EGAT and PEA do not need to be involved. With increasing deregulation, direct third-party sales may become possible in the future, which may significantly reduce PEA's business.

As part of the government's electricity sector deregulation and privatization policies, in 1992 EGAT created the Electricity Generating Company Limited (EGCO) as a wholly owned subsidiary. EGCO's main responsibility is to generate electricity and sell it to EGAT. EGAT has sold some 2,014 MW (two gas power stations) to EGCO. In 1994 EGCO was listed on the stock exchange of Thailand. EGAT has since reduced its holding to 48%. In January 1998 EGAT offered a further 15% holding for sale and received expressions of interest from around 40 companies.

---

[13] ETSU (1999b).

❑ Proposed rice husk power generation, Nakorn Pathom: EGCO is acting as the lead developer for a stand-alone power plant in Nakorn
Pathom, fuelled with rice husks from a number of surrounding mills.
Rice husks (a by-product of the rice milling process) are an agricultural
waste stream that can be used as a fuel in cogeneration plants or power
stations. About 2 kg of rice husk is required to produce 1 kWh of
electricity, but this ratio is improving as technology develops.[14] If
cogeneration is to be considered, the rice mill must process at least 5
tonnes per hour for it to be feasible at the current stage of technical development. In 1996 Thailand had fewer than 80 rice mills over this
size. If all these mills were to install efficient cogeneration technology,
around 66 MW of generating capacity could be exported to the grid.

The EGCO's proposed power plant capacity was 28 MW, with 26 MW
being exported to the PEA under a firm capacity contract, as rice is harvested all year round. Successful development of such a project could
result in profitable returns to EGCO. This would put it in a very good
position to develop similar projects in Thailand and elsewhere in the
region, providing EGCO staff with good knowledge and experience of
project development. The clear environmental benefits of this project
might also give a substantial public relations advantage to EGCO.

The key problem with this project has been securing a suitable fuel
supply, and is the main reason why it has not yet been realized. The
price paid by millers for rice deliveries is set on a short-term basis, and
it is very volatile. The millers do not deal in long-term supply contracts,
either for raw input or for processed output. Their business operates in a
completely free market, entirely supply and demand driven. Thus there
is a mismatch between the nature of their business and the long-term,
relatively fixed nature of electricity supply contracts. There are also
alternative uses for rice husks (e.g. it can be used as filler in the brick
industry, in producing charcoal from wood logs, as a fuel for domestic
purposes in villages and occasionally as bedding material for animals).
For these reasons rice millers are unwilling to be bound into long-term
supply contracts. Even if such contracts could be agreed with the mill
owners, these would be of little value, for EGCO would have little recourse if they were broken. The millers are relatively small companies
without the assets to provide compensation. If millers do not honour

---

[14] World Bank (1996).

supply contracts, it is costly and uneconomic to transport rice husks from further away.

Lessons learned   There are a number of barriers preventing the wider uptake of biomass resources for power generation in Thailand, which include the following.

• There is a lack of knowledge and awareness among mill owners and potential developers regarding the technical and economic case to support more efficient cogeneration. It is important to raise awareness among the relevant stakeholders to encourage investment in and development of efficient cogeneration systems.
• There is a lack of successful, commercial track record and experience.
• The power plants are seen by millers as complicated to run, as power generation is not their core business. Consequently, few have agreed to undertake efficient cogeneration schemes at their mills.
• Fuel supply can be difficult to secure. There is a mismatch in the short-term free-market structure of rice milling and the longer-term contracts in power supply. The PPA conditions are derived from IPP generation from gas, where longer-term fuel supply contracts can be secured. The PPA needs to be revised to accommodate the difficulties faced in contracting biomass supplies.
• It is not practical to base energy payments under the PPA on the price of biomass (as it is for fossil fuel generation) as there are regional variations in price, it is difficult to determine a base price for these fuels and there is scope for price manipulation.
• If a cogeneration plant is not integrated into a mill and the mill owner has no incentive to supply waste streams to the cogeneration plant, then there is little guarantee of biomass supply to the power plant, particularly as there is growing competition for agro-industry wastes from other users. The best thing to do is get the biomass fuel supplier bound to the project on a long-term delivery contract, or include the supplier in the power enterprise or consortium.
• There could be institutional difficulties. For example, the PEA is obliged to support the SPP scheme, but it could be detrimental to its business in several ways. Encouraging industry to increase its own generation, perhaps with the possibility of third-party sale in the future, represents a direct reduction in market share for PEA in the industrial sector, which in general pays the highest tariff and has

low distribution costs. The work incurred in strengthening the distribution system, administration, operation and maintenance, is disproportionate to the amount of power which is gained through SPPs at present.

- Supporting institutions for maintenance and spare parts do not yet exist and there is a lack of relevant specialist advice.
- Potential developers and mill owners are not aware of the support packages on offer from central government or ODA from bilateral or multilateral institutions.
- The impacts of mill size, fuel transportation and seasonality of fuel supply/electricity demand are difficult to quantify.
- There are financial barriers, particularly in respect to high-risk premiums from financiers who do not understand the technology or sector.

Case study 7: palm oil waste in Indonesia[15]

Project outline   The worldwide demand for palm oil is growing rapidly, and new plantations have been established in Indonesia and elsewhere to keep up with the demand. In 1993 Indonesia had 104 palm oil mills with a processing capacity of 3,888 tonnes of fresh fruit bunches (FFB) per hour and an estimated 1.2 million hectares of palm plantations. The largest estates are around 10,000 hectares and have mills with processing capacities of 60 tonnes FFB per hour. Prior to the recent economic and political turmoil in Indonesia, most major private-sector palm plantation owners had ambitious plans for expansion. If these plans are realized, the industry will expand vastly over the next two decades to 2020. The majority of mills at present are located on Sumatra, with a small number on Java, Kalimantan, Sulawesi and Papua Barat (formerly Irian Jaya). The government of Indonesia (GOI) through its transmigration programme encourages new plantations to be located on sparsely populated islands in Indonesia.

   Palm mills produce both solid and liquid wastes. The solid wastes are the fibres, shells and empty fruit bunches (EFB). Shells and fibres are typically combusted to generate process steam and power. EFB are incinerated, used as fertilizer or disposed of. Full fruit bunches of oil

---

[15] This case study is based on information from various sources including ETSU (1999b).

palm can be harvested year round, so the mill can be productive in all months, with a peak at the end of the dry season.

Most large mills (around 10,000 hectares, processing 60 tonnes per hr FFB) produce their own electricity and process steam using low-efficiency boiler/back pressure turbines, and rely on diesel generators for power when the mills are not operating. There is considerable potential for using more efficient and advanced boiler and power-generating technology in palm mills so that excess power can be generated and exported to the grid. The more efficient systems could be installed in existing mills or new mills. Alternatively, waste from two or more mills could be used at a purpose-built power-only plant (i.e. no process heat generated).

In the past there has been little incentive to produce surplus power for export to the grid. There is now, however, increasing interest in forestry and agro-industries generating surplus power to export to weaker areas of distribution networks, remote regional grids or mini-grids. The rural electrification plans of the GOI cannot be realized by grid extension alone, especially as Indonesia is made up of thousands of islands. The electric power supply system is operated by the state electricity corporation PT Perusahaan Listrik Negara (PLN), which since 1994 has run as an incorporated company, PT.PLN (Persero), to give it more freedom of action. PLN is now looking more favourably at stand-alone systems (e.g. solar PV) or embedded generation (e.g. power from agricultural processing mills and mini-hydro) to help achieve rural electrification. The initiative which has created a potential market for renewable energy power in Indonesia is the PSKSK, a scheme enabling long-term PPAs to be made by PLN with SPPs. This programme was initiated by ministerial decree in 1995, and the first call for proposals was issued in 1996. The PSKSK was set up for systems less than 30 MW on Java and Bali and less than 15 MW elsewhere.

To date very few PPAs have been signed under the PSKSK. This is mainly due to the Asian economic crisis that began in 1997, as a result of which negotiations with financiers and foreign developers were stalled. The slowdown in the economy and the associated reduction in energy demand mean that PLN is now in the position of having excess power generating capacity, and consequently the impetus for encouraging additional generating capacity via the PSKSK has diminished. This is reflected in the fact that since 1996 there has been no review of the tariffs for PPAs, which still stand at pre-crisis rates. Although the incentives for SPPs are currently not good, there are still valuable lessons that can be learned from the negotiations as far as they went before the crisis hit.

The following players are involved in encouraging and setting up SPPs in Indonesian agro-industry.

- *The national government.* The Directorate General of Electricity and Energy Development (DGEED) of the Ministry of Mines and Energy was involved in setting up of the PSKSK, along with other government departments such as the Department of Public Works. DGEED coordinates and supports small PPA project developments.
- *The utility.* PLN is responsible for buying power produced by SPPs and confirming that the power is needed so that PPAs can be set up.
- *Financiers.* In 1997 the World Bank approved a loan of US$66.4 million to be used to provide debt financing for renewable energy projects under the Renewable Energy Small Power (RESP) project. Project developers should be able to apply for this support via an Indonesian participating bank (four were identified). The World Bank was also involved in providing advice for the formation of the PSKSK, identification of prospective mills and feasibility studies. A Global Environment Facility (GEF) grant of US$4 million for technical assistance was also approved. Grants of up to US$100,000 per project will be available for pre-investment studies (including pre-feasibility and feasibility studies). Developers are likely to apply for such funds.
- *Developers.* Both foreign and national developers have been involved in setting up SPPs.
- *Mill owners.* These include owners of palm oil mills, sawmills, rice mills and sugar mills.

❑ Torgamba palm mill: The early stages of developing an efficient cogeneration plant at the Torgamba palm mill 400 km south of Medan in northern Sumatra have been taken as an example through which to look at the players involved and lessons learned. The mill owner is the state-owned agro-industrial company PT Perkebunan Nusantara III (PTPN3), and the developer is a small foreign company.

The mill generally operates six days a week and processes 250,000 tonnes FFB a year. On a typical operating day, the steam boilers are fired from 10.00 a.m. until all the FFB have been processed. This usually occurs in the early hours of the morning, between 2.00 a.m. and 4.00 a.m. During mill operation, the back pressure turbines are used to supply the electricity requirements of the mill and the company's offices and houses; typically the total load is 700 kW. During periods when the process boilers are shut down, the residual demand of 100–120 kW is supplied by a diesel generator.

A pre-feasibility study was carried out by the developer with data from PTPN3, and a contracting engineer carried out the preliminary design, specification and costing for the proposed cogeneration plant. The proposed plant is based on an extraction/condensing steam turbine that is estimated to give around 41.8 GWh of excess electricity per year. This is based on the assumption that all EFB, fibre and shells are used as fuel. The maximum generator capacity is 7.8 MW; however, when process steam is being extracted from the steam turbine during mill operation, the output falls by up to 3 MW. Using the design specifications provided by the contracting engineer and the estimated power price available from PLN in the future, the developer wrote outline business development plans for the project.

There are additional sources of biomass fuel, which could be used together with fuel from the host palm mill. EFB could be used from a second 60 tonne per hour palm oil mill operated by PTPN3, located 15 km from the Torgamba mill. Other potential sources include waste palm trunks and palm fronds, and inputs from other nearby mills. If these additional biomass sources were secured then alternative, larger cogeneration designs might be considered.

A memorandum of understanding (MoU) was entered into by the developer with the mill owner PTPN3. This MoU assured both parties that they would work together to pursue the project. These companies would work exclusively with one another, for a reasonable period of time, without breaching commercial confidences.

In 1997 an NGO called Winrock International approved financial support (up to a limit of 50% or US$50,000) for the developer to carry out a feasibility study at the Torgamba mill. Winrock International operates a programme to share the costs of project pre-investment studies with private developers. The programme aims to reduce the financial risk of developers and to establish commercial viability and secure financing. The feasibility study is crucial for securing financial investment for the project. It therefore has to include detailed analysis in technical, financial and legal areas. It must also include the identification of problems and risk areas, and show how these can be overcome. Because of the economic crisis in Asia, the project was halted before financial closure was achieved. However, the project is still seen as viable.

If the project is successful in developing this palm oil mill for exporting electricity to the grid at commercial or near-commercial rates, the small company will be provided with a substantial income. The incentive for the developer is that the project could also be replicated a great number of times, making the developer a world leader in this field, in a very strong commercial position. It is very important to keep the mill owner interested and committed to the project; and as an incentive the mill owner is offered a share of the equity.

Lessons learned

*   The World Bank was influential in getting the Indonesian government to establish the PSKSK, which makes it possible for SPPs to export electricity to the grid on commercial terms. The World Bank is also playing a key role in providing support to establish SPPs via the loan announced for the RESP project, identifying potential mills suitable to become SPPs and carrying out feasibility studies for selected mills. This shows the valuable role that multilateral aid agencies can play in creating an enabling environment for grid-connected renewable energy projects.
*   The Asian financial crisis beginning in 1997 brought the SPP negotiations to a halt; although these are now beginning to resume, the significant delays incurred have increased transaction costs. Smaller projects will be affected more than larger ones as the transaction costs are relatively fixed and vary little with the size of the project.
*   The conditions of the PPA are crucial as they need to encourage SPPs to set up and also give financiers enough confidence to invest in the SPPs. The tariff level in particular is important, as if it is too low SPPs will have no incentive for exporting power to the grid.
*   Security of fuel supply is a continuing risk as the agro-industries are not geared to long-term contracts and there are competing uses for the waste.
*   The strength and stability of institutions involved, such as PLN, is important to the sustainability of the project and the level of risk as perceived by financial organizations.
*   There needs to be strong commitment to the PSKSK scheme from government and the private sector for it to work.

- It is important to use proven technology wherever possible, or carry out a pilot-sized project if novel technology is the only option. Technology may need adapting for local conditions.
- Good relations between the developer and mill owner are crucial, as the developer has to secure a substantial amount of investment upfront, and must be able to trust in the mill owner to keep commercial information in confidence.
- Foreign exchange risks are very real, as has been shown by the recent Asian economic crisis. These can be minimized for the developer if power payments are linked to the US dollar, but then the currency risk is heightened for the utility, which would be hit by any unfavourable shift in exchange rates.
- The availability of loans and grants for pre-feasibility studies and feasibility studies is essential to help the market develop in the early stages. Once the market is established, then developers and commercial banks are more willing to invest in such activities without additional backing as the perceived risk is lower.

Case study 8: bagasse in India[16]

Project outline   When sugar cane is crushed in sugar mills, the waste biomass is called bagasse. Bagasse-based low-pressure cogeneration has been practised in India to meet the internal energy requirements of the sugar mills since the beginnings of sugar production in India. In 1993 the Ministry of Non-conventional Energy Sources (MNES) initiated a task force to identify the bagasse-based high-pressure cogeneration potential for export of surplus power to the grid. The task force identified the potential for and barriers to the promotion of high-pressure bagasse-based cogeneration and produced a comprehensive policy in January 1994; thereafter a national programme on bagasse-based cogeneration was formulated.

Individual state governments are not required to follow the lead of central government on power matters. The guidelines laid down by the MNES were adopted by only some of the state electricity boards (SEBs), such as the Tamil Nadu Electricity Board (TNEB) and the Karnataka Electricity Board (KEB). The various cane-growing states announced

---

[16] This case study is based on information from various sources including ETSU (1999b).

their respective policies; Table A1.1 indicates how these are implemented. Where SEBs have favourable policies on biomass cogeneration, sugar mills have adopted the practice enthusiastically, recognizing the benefits to their businesses. It was forecast that the number of mills using high pressure cogeneration might reach 20–30 by the end of 1999. The estimated potential for high-pressure bagasse cogeneration in the state of Tamil Nadu alone was 3,500 MW, of which 116 MW was operational by 1998.

Bagasse cogeneration is attractive to sugar mill owners because:

- it can make money for them;
- it enables the cane-crushing plant to be self-sufficient in heat and power;
- it represents diversification of the business;
- it solves waste disposal problems and adds value to the sugar production process;
- government-backed financial incentives are available to encourage investment, yielding short payback periods;
- in Tamil Nadu, there is a guaranteed market for power generation with the TNEB at an adequate buyback rate to meet investment criteria;
- there is a potential for offering an investment/turnkey development service to other sugar mill owners;
- in Tamil Nadu, there is the potential to supply power to sister companies (in cooperation with TNEB), thus reducing their costs;
- there is potential for sale to third-party industrial consumers at more lucrative buyback rates, should state legislation allow this in the future.

The MNES offers a range of financial incentives and grants to encourage sugar-mill owners to consider efficient high-pressure cogeneration and export of surplus power to the grid. These include:

- grants for bagasse-based cogeneration demonstration projects;
- grants towards the cost of detailed project reports (e.g. feasibility studies);
- interest subsidies through financial institutions.

The financial incentives are structured so as to encourage applications from both public- and private-sector mills, with public-sector mills receiving higher levels of support. Bagasse cogeneration is attractive to

Table A1.1: State-level incentives for cogeneration projects in India

Policy element	Maharashtra	Tamil Nadu	Karnataka	Uttar Pradesh	Madhya Pradesh	Punjab
Private participation	Allowed	Allowed	Allowed	Allowed	Allowed	Allowed
Power wheeling charge (% of energy generated)	[a]	15	6	12.5	2	2
Power banking charge (% of energy generated)		2	6	Allowed		
SEB buy-back rates (Rs/kWh)	>4 MW: 2.25 <4MW: 2.20	2.00	2.00	2.25 or, if lower, at a rate ensuring 16% return on equity	2.25	1.50
Sales to third party		Allowed	Allowed	Allowed within 5 km	Allowed	Allowed
Capital subsidies	Participation in equity up to 60% by electricity board		Rs2.5m per MW			As extended to other industries
Other concessions		Exemption from electricity generation tax for captive use	Exemption for 5 years from electricity tax for captive use		Exemption from sales tax and other concessions applicable to new industry	Exemption from duty, sales tax benefit capital subsidy

[a] Where a cell is empty, there are no state-level incentives related to that policy element in that state.

the MNES because it offers the prospect of large-scale, grid-connected renewable energy capacity at low levels of subsidy. It involves profitable companies, with the means to develop the potential at their factories; and it demonstrates how renewable energy sources can provide large amounts of power, while simultaneously solving a waste disposal problem.

In order for a sugar mill to qualify for a MNES grant for cogeneration, the relevant state government must contribute at least Rs2.5 million per MW (approximately US$56,000 per MW) of surplus power generated by the sugar mill. In some states, such as Karnataka, 'single window agencies' (one stop shops) have been established to ensure swift clearance of approvals for such power projects. All sugar-producing states allow private-sector development of cogeneration projects. State governments find bagasse cogeneration attractive because it helps reduce power deficits, it can help industrial development and it mobilizes private-sector investment in power generation.

Financial institutions are involved in bagasse cogeneration as intermediaries between the government (MNES) and the developers. They process the applications for capital subsidies (grants) and access to low-interest loans. Grants, which are given only to existing mills, may take the form of:

- a capital subsidy of up to Rs7 million per MW (approximately US$0.16 million per MW) of surplus power (subject to a maximum of Rs60 million, approximately US$1.3 million); and
- a long-term loan of up to Rs13 million per MW (approximately US$0.29 million) for surplus power at an interest rate of 9%, repayable to the financial institution after payment of any commercial loans.

Financial institutions are attracted to bagasse cogeneration projects because they can make a reasonable return on investment and the government of India is willing to subsidize the interest rates they can offer, making it easier for the financial institutions to attract additional customers.

The Tamil Nadu Energy Development Agency (TEDA) is the nodal agency for renewables in Tamil Nadu state. The agency, which reports to the energy secretary of the Tamil Nadu government, administers MNES incentives for renewables within the state. Better data than were previously available are being gathered on biomass resources in Tamil Nadu state, with particular attention being given to feasibility studies for potential generation projects. The TEDA sees the development of biomass cogeneration as a good thing as it helps to satisfy the agency's renewable energy deployment targets; it encourages the involvement of private-sector companies with capital to spend on renewable energy; and the experience gathered might help TEDA to become a more autonomous, self-financing organization.

The TNEB has worked closely with the MNES on the development of grid-connected renewable energy capacity in the state, demonstrating particular success with wind energy projects. The measured, well-planned approach taken has combined good demonstration projects, marketing incentives and pre-investment work, mostly for wind energy, but also for bagasse cogeneration. The TNEB acts as an intermediary between manufacturers and entrepreneurs, encouraging the development of bagasse cogeneration. It can offer practical help to developers, especially with planning new grid-connected generation projects and the associated power export facilities. It recognizes that renewable energy is important in the generation mix, and has a self-imposed target of 100 MW per year of additional private-sector-based bagasse cogeneration capacity in the state. The actual development rate will depend on the take-up by the private sector. Bagasse cogeneration is attractive to the TNEB because it increases the grid-connected generation capacity in the state; encourages private-sector investment in generation capacity; and enables the grid to be strengthened in rural areas.

❑ Kothari Sugars and Chemicals Ltd: The Kothari Group was founded in the late 1930s. It is involved in financial services, agro-business and petrochemicals. Kothari Sugars and Chemicals Ltd was established in 1961, and in 1998 had around 300 tonnes per day crushed sugar cane capacity at its sugar factory. It is planning to extend this crushing capacity. It has invested in a 12 MW bagasse cogeneration plant and the sugar mill also has a distillery attached to the plant, established in 1993, producing 45,000 litres per day of industrial alcohol. The company is progressive, looking actively to diversify into new technologies that benefit the business, and has a wide range of interests in both Kothari Sugars and the parent group.

Bagasse availability is seasonal. Kothari Sugars plans to expand the existing cogeneration plant by adding 10 MW of condensing turbine to allow the extension of the operating period from 270 to 330 days a year, using bagasse stored on site. The condensing boiler allows steam to be recycled and generation to continue when the sugar processing has stopped. The additional 10 MW was selected by the MNES for a demonstration project grant. The payback period calculated for the project is two years, since the boilers already exist and only the turbine has to be purchased.

Kothari previously burned bagasse to dispose of it. The company installed the bagasse cogeneration plant as a means of adding value to the sugar process by making use of a waste product. Sufficient pay-back is available from the rates offered by the TNEB to make the investment worthwhile. The TNEB would prefer continuous rather than seasonal power, so Kothari is considering extending the genera-tion period by burning lignite out of season.

The sugar industry is not growing very fast at present and world market prices are depressed. This is preventing some sugar mills from investing in bagasse cogeneration. There are also alternative uses for the bagasse such as pulp for paper factories.

The state government's declaration of appropriate policies has cre-ated an environment suitable for the promotion of cogeneration. The following main factors have made the incentive mechanism success-ful:[17]

- *favourable policy framework:* both central and state governments have adopted coordinated policies to assist the development of the bagasse cogeneration potential;
- *adequate buyback rates/financial terms to attract investment:* even though power generation is not the core business of the sugar com-panies;
- *practical help from the SEB:* good planning and provision of suffi-cient power export facilities by SEBs are key features of successful development;
- *improved efficiency and business performance:* the sugar industry, which has hitherto been using cogeneration to power the factories themselves, perceives sales of surplus power to the grid from co-generation as a way to improve the financial efficiency of the sugar production process, and hence to improve business performance;

---

[17] Taken from ETSU (1999b).

- *productive use of waste resources:* cogeneration helps the industry by making use of the waste product bagasse as a fuel, rather than simply burning it for disposal, thus converting a cost to the business into a gain;
- *diversification potential:* there is the opportunity to take up turnkey and consultancy projects if desired.

Areas where improvements could be made include:

- *third-party sale:* currently, policy on third-party sale was due to be revised in 2000. There is the possibility that third-party sales may be allowed. The industry perceives that the introduction of third-party sales would enhance the viability of projects, providing shorter payback periods and hence accelerating investment in more cogeneration capacity;
- *access to finance:* the current dip in the sugar market is making industry wary of capital investment. An upturn in the market is needed to release investment funds.

Lessons learned

- Government policies and institutional structures are very important in creating and enabling a framework conducive to the development of cogeneration activities.
- It is a great incentive for mill owners if adequate payback rates are offered and PPAs take into account the variability in power outputs from sugar mills due to the seasonality of sugar cane harvesting, not penalizing them too heavily.
- It is important to raise the awareness of mill owners of the potential financial and business performance gains of cogeneration.
- It is important to raise awareness of the benefits created by disposing of waste streams from mills in an environmentally sound way and at the same time generating power.

Annex 2

Analysis of Case Studies: Options for Overcoming Barriers

Below are two tables showing suggested options for overcoming the barriers to the introduction of renewable energy technology identified in the case studies in Annex 1. These should be regarded as guidelines only, as in each case the unique local and national situation, involving elements such as policy, legislation, environment and culture, will impact greatly on the success of projects.

It can be seen that some of the barriers and options for overcoming them are similar for SHS and biomass cogeneration. These barriers include:

- lack of government plans and targets;
- inappropriate fiscal policies;
- lack of support mechanisms;
- inconsistent government policies and poor communication between departments;
- lack of focus and ownership for renewable energy development;
- lack of access to information;
- lack of local skilled labour and capabilities;
- lack of information exchange;
- unclear IPR law and weak legal systems;
- lack of technical standards and quality;
- lack of confidence and knowledge;
- ineffective institutional structures;
- lack of supporting infrastructure.

Technical barriers are obviously different for SHS and biomass cogeneration; there are also some other non-technical barriers that are specific to one or other technology, depending on grid-connected status and the types of customers or end users. These include:

- lack of access to the grid for export of surplus power from cogeneration;

- vested interests restricting market access and regulations for PPAs;
- willingness and ability of users to pay for SHS;
- lack of micro-financing schemes available to households for SHS;
- social (e.g. not understanding or meeting the requirements of the household);
- small scale of SHS projects (e.g. little scope for economies of scale in manufacture);
- small size of related organizations (e.g. little capacity to install and maintain systems in dispersed locations).

Table A2.1: Overcoming barriers to solar home system projects

Barrier	Example	Options for overcoming the barrier
National policies and programmes		
Lack of clear government plans and targets	No clear government targets for solar PV capacity in the future	The developing country government should consider setting actionable policies and targets for installation of PV capacity in the future, particularly with respect to SHS for rural electrification. This sends a positive signal to developers and financiers
Inappropriate fiscal policies	High duty and VAT on imported SHS components (up to 30–40% in some cases) as compared to zero import duty on conventional grid electrification technology	If governments wish to encourage the use of sustainable energy systems, they could remove or reduce import duty and tax on SHS components, particularly batteries, PV panels, controllers and energy-efficient DC lights. They could also consider reducing import duty and tax on DC appliances that can be used in conjunction with SHS (e.g. black and white TV, radio/cassette player). Reducing import duty would not be necessary if SHS or DC components were made in-country to a high enough standard
Lack of support mechanisms	Grants and soft loans	Government-funded grants and soft loans are preferable to international financing as there is no exchange risk. However, the funds available from international sources are often much bigger. Ways to mobilize government funds include levies on fossil fuels or electricity. For example, the Energy Conservation fund (ENCON) in Thailand is collected by a levy on transport fuels
Grid electrification plans	Changing grid electrification plans make it difficult to plan SHS installations effectively	It is important for future grid electrification plans to be declared for up to the next 5–10 years and adhered to. This achieves three things: it helps developers plan the most appropriate sites to install SHS; it gives developers confidence that they will have a market for the next 5–10 years before the grid arrives; and it gives investors confidence that the market will be sustained long enough for the loans to be repaid
Lack of integrated planning for energy and development	Often sustainable energy systems are not planned at the same time as new schools, health centres etc., so opportunities for PV installation are delayed or missed altogether	If energy systems are planned at the same time as new schools and health centres, etc., then PV energy services could be designed and utilized more effectively, providing sustainable energy systems in rural communities, and expanding the market for PV-based services. It would help planning if grid electrification plans in the future were clear and also if costs compared to alternatives such as diesel or renewables were clear so

Table A2.1: continued

Barrier	Example	Options for overcoming the barrier
		that informed choices could be made as to the most cost-effective and appropriate power source
Lack of consistent policy	Increased political risk and uncertainty can raise transaction costs. Transaction costs are related more closely to negotiation time than to project size	Clear and straightforward planning regulations are needed to reduce the time it takes to process planning applications. It is also helpful to reduce the number of government departments involved in the approval process. The longer it takes to negotiate a project, the higher the transaction costs. As PV projects are often relatively small, the transaction costs are a significant burden. Small-scale projects are to be fast-tracked under the CDM (see Section 2.1.2, part ii)
Lack of focus and ownership for renewable energy development	Several government departments involved in RE schemes	It is important to have a 'champion' government department in charge of coordinating RE projects. It should be responsible for meeting agreed RE targets, and its performance should be measured against those targets

Information exchange, education and technical training

Lack of access to information	If users are unaware of the services which can be provided by SHS, there is no market for the systems. If potential developers and installers are unaware of the potential market, they do not participate in market development. If investors/financiers are unfamiliar with the technology, they are not confident enough to provide loans	It is essential to have accurate detailed resource assessments done showing the insolation regime. It is also important to carry out market assessments and disseminate this information, to encourage potential developers and financiers to develop and invest in the market. ODA can help the developing country governments by providing experts to train local staff in resource assessments and market assessments. The developing country government needs to collect together useful information on PV (and other RE sources) in a central place and take action to disseminate this to users, developers, installers and financiers. The methods for dissemination and form of information need to be chosen carefully to be most effective. For example, many people in remote rural areas may not be literate, so verbal dissemination via community meetings, or by radio or cartoon-style printed information may be more appropriate than literature. Awareness raising in schools can give children a good basic knowledge, which can be built on as they get older

Table A2.1: continued

Barrier	Example	Options for overcoming the barrier
Lack of skilled labour and capabilities	Lack of locally trained people in installation, operation and maintenance in remote areas due to lack of education regarding new technology (both PV systems and direct current appliances)	Developing country governments should consider introducing relevant training in schools and universities on direct current technology for household electricity. PV demonstration units in schools (even in electrified regions) will play a promotional role if people understand what they can do with minimal amounts of electricity. Developers should consider training local people to install and maintain equipment. It can be good to train women in maintenance skills as they may be less likely to migrate to urban areas. Men often take their new-found skills to urban areas for better paid jobs. Also children are more likely to learn the skills taught to their mother as they spend more time with them. But these two things depend heavily on the local culture and the jobs available locally and in urban areas
Lack of exchange in ideas and experiences	Valuable knowledge and experience are being created through demonstration projects, but this is not being disseminated effectively to enable people to learn from others' experiences	Many of the SHS demonstration projects have provided useful lessons on how to overcome barriers. It is important to share this information so that others can benefit from these experiences. This information needs to be collected together, analysed and disseminated to key players in an easily digestible form. Also, information on who the key players are in the developing countries, who the technology manufacturers are and what technology is available is very useful. An exchange of information between developed and developing countries is needed, along with exchange of information between developing countries. Renewable energy information networks have been established in some countries and regions. It would be most effective to get these networks to cooperate with one another and pool information and resources. There is also a need to take steps to bring information out of these networks into the more widely accessible media such as newspapers, radio and TV. The wider audience needs to be kept informed

Intellectual property and standards

Unclear law on intellectual property rights	Manufacturers are deterred from setting up joint ventures (JVs) for domestic manufacture of PV technology as they are worried their technology designs will be stolen	Setting up joint ventures to manufacture technology in developing countries brings down costs significantly, mainly because labour and some materials are cheaper in developing countries. It is therefore important to encourage local manufacture of SHS components and also DC appliances to use with the systems. However, it is not necessarily appropriate to manufacture high-tech components (such as PV cells) in devel-

Table A2.1: continued

Barrier	Example	Options for overcoming the barrier
		oping countries, as the technology is still developing and once a manufacturing plant is installed it is likely quickly to become out of date. The encapsulation of solar PV cells into panels and the manufacture of wiring, controllers, batteries, lights, ballasts and other DC appliances are, however, possible as long as there is already a reasonably developed industrial base. (For example, Thailand has good industrial capacity and capabilities, but the island states of Kiribati and Tuvalu do not.) To encourage the skills and technology manufacture to be transferred to developing countries, governments have to ensure that the companies that own the technology are confident in national law regarding intellectual property rights and the legal system in that country
Lack of supporting legal institutions	Lack of confidence that legal disputes can be resolved satisfactorily in developing countries	Developers, installers, manufacturers and financiers need to be given confidence that if there are any legal disputes (e.g. over IPRs, contracts etc.) that the legal institutions in the developing country are strong enough to deal with the dispute and they will be able to recover their costs where eligible
Lack of technical standards and quality control	A lack of standards leaves users unable to differentiate between good and poor-quality systems. Poor-quality technology is more likely to fail. Manufacturers cannot give standard guarantees for locally manufactured technology if the quality is poor; this deters JVs with manufacturers fearing damage to their reputation	Users need to be able to choose technology on performance as well as price. Therefore, it is important to have technical standards against which performance can be measured and the quality of the technology assessed. It is essential not to dictate the specification of technology which is allowed to be imported and sold too strictly, as this could destroy the local market which can only afford cheap, low-rated systems. For example, in Kenya the market has grown due to demand for cheap imports of low-rated PV panels. Developing country governments should support the adoption and enforcement of clear standards as a guideline for users, installers and developers. ODA could be used to assist developing countries in the task of setting standards. International standards should be adopted if possible so that technology exported is accepted by other countries. International standards should account for different climates where possible (dry, continental, humid, tropical etc.)

Financing

Lack of access to capital	Lack of micro-financing packages available to	It is reassuring for financiers to have local government backing for loan applications by households, to confirm

Table A2.1: continued

Barrier	Example	Options for overcoming the barrier
	households. Difficulties in convincing banks of creditworthiness and reliability in repaying loans	they are not already in debt and live at the address intended for SHS installation. It is also reassuring for financiers if a well-respected and effective local cooperative or union (e.g. the VWU in Vietnam) is involved in collecting the monthly fees and can put some social pressure on households to keep up to date with payments. Loan guarantees provided by the developer are effective in reducing the risk to the bank of non-payment, thus increasing confidence to provide the loan (this is being done successfully by SELCO in Vietnam). It is important to identify to whom in the household it is best to make the loan. For example, lending to women has shown better repayment rates in India and Bangladesh
	Lack of access to capital for Installers and developers	It is difficult for new developers and installers to get loans, as they do not have a sufficient track record to show they are creditworthy. Some form of loan guarantee from national government or ODA in the form of a grant would be useful in securing such loans. It would also be helpful if lending bodies were to recognize the PV panels as collateral for the loans, as they have a resale value. Some banks fear they will be stuck with PV panels that nobody wants, but PV module manufacturers could help by guaranteeing a certain value of the module depreciated over the period of the loan. It should be a requirement that PV projects are sustainable, e.g. that a broken lamp or controller can be replaced by locally available technology or a cost-competitive import. All installed systems would benefit from that flexibility
Lack of investment	Lack of investment prevents the SHS market from developing	Developing country governments can invest in SHS demonstration projects by providing grants and low-interest loans. Bilateral and multilateral aid organizations can invest in SHS by providing grants and loans to developers, installers and other key players via developing country national banks. It is better for loans to be received and paid back in national currency, as this removes the risk of loss due to exchange rate fluctuations. However, national sources of funding are restricted and foreign investment is growing in importance, as outlined in Chapter 3, Section 2
		Financial institutions need to revise their project assessment methodology to take account of the special features of renewable energy projects. For example,

Table A2.1: continued

Barrier	Example	Options for overcoming the barrier
		renewable energy projects typically have a relatively high up-front cost, then low operation and maintenance costs as the fuel is often free (except for biomass in some cases). Bilateral or multilateral assistance could be provided to train financial institutions in best practices for renewable energy project finance schemes, including project assessment and the terms and conditions for loans to developers, installers and users (this is being done in the Development Bank of the Philippines)
		Investment in education and capacity building to develop much-needed skills nationally and locally is also very important. Developing country governments have a role to play in setting up education facilities, and bilateral and multilateral aid has a role to play in investing in the transfer of knowledge and skills (e.g. training of trainers etc.)
		Investment is needed in SMEs to help them expand their business capabilities and supporting infrastructure to assist renewable energy technology deployment (see Chapter 3, Section 3.1)
	Tied aid projects	Tied aid can lead to the use of inappropriate foreign technology which is unsustainable economically (too expensive to replace broken parts) and practically (specific spare parts for that system unavailable or take a very long time to arrive)
Inappropriate subsidies	Subsidy of fossil fuels and rural electrification via grid extension make it difficult for SHS to compete economically	The continued subsidizing of grid extension schemes cannot be sustained in developing countries. Much of the time the true cost of rural electrification via grid extension is not apparent as costs are not transparent in national utilities. It is important that if subsidies have to be introduced they are sustainable or can be reduced to a sustainable level in the near future. 'Smart' subsidies are needed that are targeted, pro-poor, transparent and time limited. Many developing countries subsidize fossil fuels, which is a big burden on the exchequer, and richer families benefit more than poorer families who use less energy. Privatization of the electricity supply industry (ESI) in most cases encourages more economic and efficient power production and transmission systems, and can bring down electricity supply costs. The World Bank is instrumental in helping developing countries go through the process

Table A2.1: continued

Barrier	Example	Options for overcoming the barrier
		of privatizing and liberalizing their ESIs. The reduction or removal of subsidies for grid extension is allowing SHS and other renewable energy stand-alone systems to become more cost-competitive, particularly in remote rural areas
Small scale of systems	SHS are small in comparison to conventional power projects	Financial institutions are not geared to small power project loans. National banks and other financial institutions need training in best practice for renewable energy project appraisal, taking into account the small scale of some projects
Small size of organizations	As the PV market is relatively small, many of the players are SMEs. However, there are a few big organizations, such as Shell and BP, which are expanding their solar capabilities through mergers and acquisitions	The price of PV technology is still relatively high and an increase in the scale of manufacture is needed to reduce prices. The merger of manufacturers helps to improve the technology and reduce manufacturing costs. However, it can also have the effect of putting SMEs out of business, removing from the marketplace the cheaper low-capacity, lower-quality technology which can play an important role in establishing the PV market, as seen in Kenya. So it is important to help SMEs survive. To assist SMEs the speed of processing project applications can be reduced by streamlining and allowing one government department to deal with the procedure from beginning to end. This would reduce the transaction costs which can be crippling for small organizations. SMEs need help with investment if they are to grow and keep up with the expanding market. Many of the existing local organizations are too small to cope with projects on a large scale, and the supporting infrastructure is not there. To overcome this, projects can be installed over several years to allow time for capacity building and infrastructure development to take place at the same time

Other

Inefficient institutional structures	Restricted power supply rights	The legal right to supply power may be limited (e.g. in the Philippines the local electricity board has sole rights to supply electricity on each island). This discourages developers from trying to install SHS. Such restrictions should be reviewed and franchise arrangements made easy to negotiate and set up
	Fixed price for power	If the price for the sale of electricity is set at a low level by government, this can deter RESCOs from setting up and supplying electricity services (e.g. in the Philippines, the government has set a maximum price for electricity supply

Table A2.1: continued

Barrier	Example	Options for overcoming the barrier
		in the small islands, which is equivalent to the price of grid electricity on the main islands; this fixed price is too low to make it economical for developers to install SHS). Governments should review any such policies and allow RESCOs to charge tariffs that allow them to recover their costs fully and to invest in the expansion of services. If the RESCOs try to charge too much, people simply will not be able to afford their services and the market will automatically regulate itself
	Lack of coordination and cooperation between government departments	As mentioned above, the planning and implementation of projects are frequently spread across different government departments. It would be more effective to focus activities on one department and make it a 'champion' for renewable energy. An alternative would be to review the communications and cooperation between departments to see if these can be streamlined, and give them RE targets to achieve jointly against which they will be measured
Lack of supporting infrastructure	Lack of access to spare parts and people trained in maintenance in remote rural areas	It is essential to build supporting networks based on existing infrastructure and businesses as the nascent PV market is too small to provide enough business for such services to survive on their own. It is important to increase the number of SHS in any region to a large enough number to make the supporting infrastructure economic to run, and SHS projects sustainable. For example, the SEC in Kiribati at present maintains 250 SHS funded by the European Commission. This number needs to be increased to over 1,000 in order for the payments on systems to cover the staff maintenance costs, replacement parts and overheads of the organization. If this number of systems is reached then even the panels can be replaced after 20 years, making the project self-sustaining financially
Vested interests	PV electrification is still seen as a competitive rather than complementary option to conventional electrification by some utilities	Vested interests concern not only the utilities, but everyone involved in all aspects of the energy business, e.g. sale of gensets (diesel, gas), sale and transportation of fuel, and maintenance of generation and distribution equipment. Therefore awareness must be raised of the new business opportunities and potential for job creation offered by the supporting infrastructure for PV systems, e.g. the retail supply and service network
Variable ability to pay	Irregular or non-existent cash income. Remote communities may trade with commodities not cash	When loans are being set up, the ability to make regular cash payments should be assessed. Financiers need to take this information into consideration when drawing up repayment terms. They should consider the seasonal fluctuations in cash flows related to the sale of livestock and

Table A2.1: continued

Barrier	Example	Options for overcoming the barrier
		harvested crops. More conveniently timed payments should be agreed if possible
		Very remote communities may more commonly trade in commodities than cash. Some research needs to be done on how these communities might best arrange payments to the financiers and if the financiers could accept payment in some form of commodities or labour. A demonstration or pilot financing scheme based on commodity payments is needed to see how best this might work. This will be influenced greatly by the local circumstances and culture
Variable willingness to pay	Poor monthly fee collection rates	It is important to do a detailed survey before SHS are installed to assess the willingness and ability of households to pay for the systems. Payment collection schemes should be designed to take into consideration any relevant local factors (e.g. density of population, local culture, regularity of cash income) which might influence the most effective method of fee collection. In Kiribati, the local technician visits households each month for maintenance and fee collection, which is quite effective. In Vietnam, where the VWU collects fees, social pressure encourages regular payment
		It is important to back up fee collection schemes with a strict system removal policy when households are, for example, three or more monthly payments in arrears. If this is not done then the households and their neighbours will see they can still have the system without paying, and problems with fee collection will escalate
Lack of confidence	Users may have seen past SHS fail so are unwilling to take out loans to purchase them	It is important to build up the confidence of users by implementing well-run SHS pilot projects and disseminating the results of these projects. It is equally important to analyse failed projects and explain why they were unsuccessful, and reassure people that the mistakes made will be addressed and avoided in future projects
	Guarantees	Provision of technology guarantees also helps build the confidence of users, so it is important to make sure the guarantees are upheld. This is easier to do when systems are installed as part of a larger project, as suppliers to bulk purchase contracts will have had to supply guarantees on the equipment as part of the competitive tendering process. The project coordinator should be able to chase up the supplier to ensure replacement of failed equipment. It is more difficult to chase up guarantees on individual systems bought direct by households through a dealer or installer

Table A2.1: continued

Barrier	Example	Options for overcoming the barrier
Social factors	The decision-maker in the household may not be the one who would benefit most from the services provided by the SHS	It is important when marketing the SHS to understand who in the household makes the decisions and who controls the finances. The local culture will impact on this greatly. SHS can benefit everyone in the household, but some might benefit more than others. If the man is the decision-maker in the household, the benefits which must be highlighted to him include access to TV, video, radio, good lighting for reading and eating, as the fact that his wife would have better lighting for cooking, tending to the children and sewing/handicrafts, etc. may not be of so much interest to him. The system must be designed to meet the requirements of the users. This is best done with a high level of consultation with potential users on what services they require from the system and where in the house they want the services
Technical factors	Unfamiliarity with local environmental conditions, e.g. local levels of insolation, appropriate system positioning, orientation and angle, proximity to fast-growing vegetation, corrosive marine environment, etc.	The technology and its installation need to take into consideration local conditions. Local insolation levels need to be measured to size the systems correctly against the requirements of the household. The geographical location of the site has an impact on the orientation and angle of panel mounting. If the site is near the sea, wiring needs to be installed to prevent corrosion where possible. Technology needs to be easily maintained by field technicians or households (i.e. keeping delicate circuitry and the need for tools to a minimum)
		Some basic maintenance skills and knowledge must be taught to the household and local technician, including: keeping the panels free from shading by vegetation; cleaning the panels regularly to remove dust and dirt; using clean water (preferably distilled water) to top up the batteries; checking the battery regularly with a hydrometer to catch any damage to cells early enough to reverse the process; installing systems with the minimum amount of wiring and connections to reduce system losses; making wire connections with proper connecting blocks, not just twisted wires, to reduce system losses; mounting panels securely to withstand storms and high winds
		There are many elements of good practice for system design, installation, use and maintenance of the systems. Users, installers and maintenance technicians should be given regular training to keep their skills updated, particularly when systems start to get older and skills in diagnostics and more difficult maintenance are required

Table A2.1: continued

Barrier	Example	Options for overcoming the barrier
	Misuse of systems (disconnected controller, charging additional batteries)	It is important to educate users in the proper use of and care for their systems, to try to prevent misuse and damage. There are differing views as to whether misuse is reduced if the system belongs to the user or remains the property of a RESCO. If the RESCO owns the technology it can try to minimize misuse with unannounced spot checks and a strict system removal policy if abuse is detected

Table A2.2: Overcoming barriers to biomass cogeneration projects

Barrier	Example	Options for overcoming the barrier
National policies and programmes		
Lack of clear government plans and targets	No clear government policy to encourage energy from biomass projects, or any targets for biomass cogeneration	Clear government targets and policies to encourage biomass cogeneration and purchase of surplus electricity would encourage mill owners and developers to install biomass cogeneration technology, as it would give them confidence that there is a market for the electricity and government would support the development. A good example is cogeneration from bagasse in India. Local government has set targets for cogeneration and the local electricity supply industry is buying the power at rates that make it attractive for sugar millers
Inappropriate fiscal policies and support mechanisms	Lack of financial incentives	Governments often do not see the merit in small power generation systems as they are not seen to be cost-competitive with larger-scale thermal plants. They should recognize the externalities of power generation from fossil fuels and the benefits of generating energy from waste, and compare fuels on a life-cycle basis with externalities internalized. This may encourage them to take action such as providing grants and soft loans to mill owners for demonstration biomass cogeneration systems, or reducing import duty and taxes on cogeneration technology to lower the investment cost to the mill owner
Lack of access to the grid	Millers unable to sell electricity to the grid	Millers are not likely to make use of crop residues to generate spare electricity unless legislation allows them to export it to the grid or to supply power direct to customers (industrial or domestic) at an attractive tariff. The ESI needs to set up PPAs with the mill owners to arrange export to the grid. The terms and conditions of the PPAs are very important if mill owners are to be persuaded to export power. The possible uncertainty of biomass supplies could make it difficult for mills to export a firm power capacity, and the ESIs often pay a lower tariff for variable power export. The buyback rates offered under the PPA must make it economically worthwhile for the mill owners to install the systems. Off-peak/peak contracts could be offered to try to match supply with demand

Table A2.2: continued

Barrier	Example	Options for overcoming the barrier
Lack of consistent policy	Lack of provision to sell electricity direct to third parties	Although developing countries are starting to privatize and liberalize their ESIs, not many allow direct sales to third parties. Direct sales would encourage more mill owners to install cogeneration, as they would get a better price for their electricity than under the PPA
Lack of focus and ownership for renewable energy development	Many government departments involved in the planning and approval of cogeneration applications	If governments want to encourage cogeneration installations, they need to make the planning and approval process quick and easy, or transaction costs could be too high and become a deterrent to mill owners and developers. In Karnataka state in India the government has set up a 'single window' agency system for bagasse cogeneration planning applications to ensure they get processed quickly

Information exchange, education and technical training

Barrier	Example	Options for overcoming the barrier
Lack of access to information	Agro-industry mill owners not aware of the benefits (economic and environmental) cogeneration can bring	Government market assessments and feasibility studies in India have shown that if mill owners are aware of the benefits to their industry then they are keen to install the cogeneration equipment. Carrying out and disseminating the results of feasibility studies and demonstration projects is a good way to raise awareness of the potential for biomass cogeneration. National government should consider funding relevant feasibility studies and demonstration plants. Existing national, regional and international RE information networks should be encouraged to work in collaboration with one another and pool information to be disseminated
Lack of focus in the biomass industry	Power generation is not a core business for mill owners	Mill owners are not traditionally power project developers and will not focus on the optimal system efficiency for power generation. Instead they will focus on the optimal efficiency for agricultural processing and waste minimization. In cases where the mill owners are not willing to consider power generation, there could be some merit in an ESCO collaborating with several mills and utilizing their waste biomass

Table A2.2: continued

Barrier	Example	Options for overcoming the barrier
Lack of skilled labour and capabilities	Lack of trained engineers	Developing country governments need to invest in training and capacity building. For example, specific training on cogeneration technology could be included in university courses
	Mill owners not knowledgeable about cogeneration technology and power generation dismiss it as 'too complicated'	Mill owners need help in planning, installing and O&M of the cogeneration technology. In some states in India, the electricity boards help sugar mills plan and install the system that will export power to the grid. Mill owners or developers need to employ experts to install the cogeneration plant and qualified engineers to operate and maintain it. The mill owner will have problems maintaining the equipment if there are no locally trained and available engineers
Lack of exchange in ideas and experiences	Lack of analysis and dissemination of experiences with biomass cogeneration projects	Biomass cogeneration and export of excess power to the grid is relatively new. There have been demonstration projects in different developing countries with a range of biomass resources resulting in variable success and experiences. In some cases demonstration projects have yet to be realized due to difficulties in negotiations, or the Asian economic crisis bringing a halt to developments. It is vital to disseminate lessons learned from these demonstration projects to allow people to learn from others' experiences. Developing country governments and ODA can play a key role in this

Intellectual property and standards

Barrier	Example	Options for overcoming the barrier
Lack of supporting legal institutions	Lack of confidence that legal disputes can be resolved satisfactorily in developing countries	Developers, manufacturers and financiers need to be given confidence that if there are any legal disputes (e.g. over IPRs, contracts etc.), the legal institutions in the developing country are strong enough to deal with the dispute and they will be able to recover their costs if appropriate

Financing

Barrier	Example	Options for overcoming the barrier
Lack of access to credit	Financial institutions not used to project appraisal for small-scale power systems or renewable energy systems	Financial institutions need training in best practice methods for renewable energy power project appraisal. Multilateral organizations and NGOs are well placed to offer this type of assistance. Mill owners have the advantage that they have the collateral of their mills to help secure a loan

Table A2.2: continued

Barrier	Example	Options for overcoming the barrier
Mill owners not willing to take the investment risk	Mill owners not willing to take out loans to invest in cogeneration technology	Mill owners will be willing to take the risk of investment if the legal framework is sound and the market conditions are improved, i.e. the SPP tariff and agreement conditions are acceptable. The Indian bagasse case study shows that when the conditions are right they are willing to invest
Inappropriate subsidies	Subsidies to the ESI undermine the price received under PPAs	The ESI in many developing countries is subsidized, and has hidden costs for power generation and grid extension. This is not a sustainable situation, and if RE systems are to compete with fossil fuel generation economically in the future, there needs to be a reduction or removal of subsidies for power generation, allowing RE systems to become more cost-competitive
Other		
Insecurity of biomass resources	Difficulty of securing long-term biomass contracts as the main activity of the mills is to process crops, so mills work when crops are harvested and may stand idle at other times	There is a mismatch between the short-term nature of biomass markets and the long-term nature of power supply contracts. It is difficult, for example, to get rice-mill owners to sign a long-term contract to supply rice husks, as they may choose to close their mill temporarily when rice prices are down, and there are other competing uses for the rice husk. Also, some biomass crops are more seasonal than others (rice has 1–3 harvests per year, sugar cane 1–2 harvests per year). This leaves developers unable to predict their firm capacity. If cogeneration is attached to a specific mill, and the mill owner is invited to take a share in the equity of the cogeneration system, the owner is more committed to supplying the rice husk for cogeneration. It is better not to have to transport biomass resources as this becomes uneconomic. Mill owners need to be convinced of the economic and environmental advantage of cogeneration. This is best done with demonstration projects and feasibility studies by developing country governments and ODA
Technical factors	Some novel technology is still being developed	For some biomass resources such as palm oil EFB and bagasse, combustion technology is reasonably well developed. But for combustion of rice husks, or the anaerobic digestion of liquid wastes from crop residues, the technology is more recent, and has not

Table A2.2: continued

Barrier	Example	Options for overcoming the barrier
		been proven substantially in the field. This makes mill owners, developers and financiers a little apprehensive about investing in such technologies. Demonstration projects are needed to remove this apprehension. It is best to use proven technology where possible as novel technology can raise many unforeseen problems
Ineffective institutional structures	Lack of coordination and cooperation between government departments	As mentioned above, the planning and implementation of projects are frequently spread across different government departments. It would be more effective to focus activities on one department and make it a 'champion' for renewable energy. If this is not possible, communications and cooperation between departments should be reviewed to see if this can be streamlined; they should be given RE targets to achieve jointly against which they will be measured
Lack of supporting infrastructure	Lack of access to trained engineers	Mill owners will probably need to employ additional staff to operate the cogeneration system, including an appropriately trained engineer. It may be difficult to find a locally based engineer or one who is willing to move to the area on a long-term basis. Power-engineering training courses in universities could include specific course units on biomass cogeneration to ensure new engineers are kept up to date with developments in technology
Vested interests	Competition with existing generators	Generation and export of power to the grid could be in direct competition with the local ESI in the future if cogeneration grows substantially in capacity. At present mills are not allowed to supply power direct to third parties as this would be taking away the ESI's most profitable supply contracts. It is not yet clear if third-party supply will be allowed, but there are obvious vested interests in the electricity supply industry that would pose obstacles to this happening

References

Allen, M. (2000), 'Distributed Energy Financing'. Presentation at Village Power 2000, Washington DC, 4–8 December.

Begg, K. G., Parkinson, S. D., Mulugetta, Y., Wilkinson, R., Doig A. and Anderson, T. (2000), *Initial Evaluation of CDM Type Projects in Developing Countries: Final Report and Annexes*. A research project funded by the Department for International Development and coordinated by the Centre for Environmental Strategy, University of Surrey. March.

Berkovski, B. (1995), *World Solar Summit Process*. Proceedings of Workshop on Financing the Development and Deployment of Renewable Energy Technologies. Oak Ridge Institute for Science and Education for the US Department of Energy, contract DE-AC05–760R00033. May.

Bernow, S., Kartha, S., Lazarus, M. and Page T. (2000), *Cleaner Generation, Free Riders and Environmental Integrity: Clean Development Mechanism and the Power Sector*. An analysis for the World Wildlife Fund by the Tellus Institute and Stockholm Environment Institute–Boston Centre. September.

Cabraal, A., Cosgrove-Davies, M. and Schaeffer, L. (1996), *Best Practices for Photovoltaic Household Electrification Programs: Lessons from Experiences in Selected Countries*. World Bank Technical Paper No. 324. Asia Technical Department Series.

Cawood W. N. (2001), 'Maphephetheni Renewables at Work: How Renewable Energy Can Be Applied to Create Jobs and Generate Income in Rural Communities'. *Renewable Energy World*, May–June.

CEC, DGXVII (1997), ATLAS Energy Technology Study 1997. Information from web pages <http://europa.eu.int/en/comm/dg17/atlas/html>.

Dixon T. (1999), AEA Technology Environment (ETSU), *Technology Transfer and Exports Promotion Strategy for the DTI's Clean Coal Technology Programme: 1999–2002*. Discussion paper prepared on behalf of the Department of Trade and Industry, UK, for discussion by the Advisory Committee on Clean Coal Technology.

E&Co. (2000), 'Meeting the Unmet Demand for Modern Energy'. Draft paper by E&Co., Bloomfield, NJ, October.

ETSU (1999a), *Integration of Renewables into Energy Systems*. A report produced for the Department for International Development. Part I, 'Main Report'. AEA Technology Environment (ETSU).

ETSU (1999b), *Integration of Renewables into Energy Systems*. A report produced for the Department for International Development. Part II, 'Case Studies and Workshop Proceedings'. AEA Technology Environment (ETSU).

FAO (1999), *Forest Energy Forum*, No. 5, December. Rome: Food and Agricultural Organization of the United Nations.

FAO (2000), *The Energy and Agriculture Nexus: Environment and Natural Resources*, Working Paper No. 4. Rome: Food and Agriculture Organization of the United Nations.

Forsyth, T. (1999), *International Investment and Climate Change: Energy Technologies for Developing Countries*. London: RIIA/Earthscan.

G8 RETF (2001), *Report of the G8 Renewable Energy Task Force*. July. <http// www.renewabletaskforce.org>.

GEF (2001), *New Business: Renewable Energy, Geothermal, Biomass, Wind, Fuel Cells and Solar*. Global Environment Facility Booklet. Washington: GEF, March.

Gillett, W. and Wilkins, G. T. (1999), AEA Technology Environment (ETSU), *Evaluation of the PREP Component: PV Systems for Rural Electrification in Kiribati and Tuvalu*. Final Report for the European Commission, DGVIII (Development). Didcot: AEAT.

Goldemberg, José (1992), 'Can the Environmental Costs of Industrial Development be Leapfrogged through Transfer of Technology?' Presentation to symposium on Free Trade and the Environment in Latin America. Loyola Law School, Los Angeles, California, February.

Gregory, J., Silveira, S., Derrick, A., Cowley, P., Allinson, C. and Paish, O. (1997), *Financing Renewable Energy Projects: A Guide for Development Workers*. London: Intermediate Technology Publications in association with the Stockholm Environment Institute.

Grubb, M. and Vrolijk, C. (1998), *The Kyoto Protocol: Specific Commitments and Flexibility Instruments*. Climate Change Briefing Paper No. 11. London: Royal Institute of International Affairs, Energy and Environment Programme.

Grubb, M., Vrolijk, C. and Brack, D. (1999), *The Kyoto Protocol: A Guide and Assessment*. London: Royal Institute of International Affairs/Earthscan.

Gunaratne, L. (1999), 'Challenges for a New Millennium: Solar Energy Business in the Developing World'. *Renewable Energy World* 2 (4), pp. 80–5.

ICC (1998), *Background Note: The Financial Crisis in Asia. Analysis and Summary Tables*. Washington DC: International Chamber of Commerce.

IEA (1997), *IEA Energy Technology R&D Statistics*. Paris: International Energy Agency.

IPCC (1996a), 'Summary for Policymakers'. In *Impacts, Adaptation and Mitigation of Climate Change: Scientific–Technical Analyses*. Contribution of Working Group II to the Second Assessment Report of the Intergovernmental Panel on Climate Change, ed. R. T. Watson, M. C. Zinyowera and R. H. Moss. Cambridge and New York: Cambridge University Press.

IPCC (1996b), 'Energy Supply Mitigation Options'. In *Impacts, Adaptation and Mitigation of Climate Change: Scientific–Technical Analyses*. Contribution of Working Group II to the Second Assessment Report of the Intergovernmental Panel on Climate Change, ed. R. T. Watson, M. C. Zinyowera and R. H. Moss, ch. 19. Cambridge and New York: Cambridge University Press.

IPCC (1998), *Special Report on Technology Transfer*, ch. 6. Draft 0. October.

IPCC (1999), *Special Report on Technology Transfer*, chs 4, 11. Draft 1.0, circulated for expert review. December 1998, revised March 1999.

IPCC (2000), *Methodological and Technological Issues in Technology Transfer*. Special Report of IPCC Working Group III. USA: Cambridge University Press/World Meteorological Organization/United Nations Environment Programme. Cambridge and New York: Cambridge University Press.

IPCC (2001), 'Summary for Policymakers'. Working Group I of the Intergovernmental Panel on Climate Change, January.

Juma, C. (1999), *Intellectual Property Rights and Globalisation: Implications for Developing Countries*. Science, Technology and Innovation Discussion Paper No. 4. Cambridge, MA: Centre for International Development, Harvard University. Paper available on <http:// www.cid.harvard.edu/cidtech/home.htm>.

Kaufman, S., with contributions from Duke, R., Hansen, R., Rogers, J., Schwarts, R., and Trexler, M. (2000), *Rural Electrification with Solar Energy as a Climate Protection Strategy*. Renewable Energy Policy Project (REPP), Research Report No. 9, January. The report can be found on <www.repp.org>.

Martinot, E. (2001), 'Renewable Energy Investment by the World Bank'. *Energy Policy* 29, pp. 689–99.

Martinot, E. and McDoom, O. (2000), *Promoting Energy Efficiency and Renewable Energy: GEF Climate Change Projects and Impacts*. Washington DC: Global Environment Facility. June.

Martinot, E., Ramankutty, R. and Rittner, F. (2000), 'The GEF Solar PV Portfolio: Emerging Experience and Lessons'. Monitoring and Evaluation Working Paper 2. August.

Maurer, C. and Bhandari, R. (2000), 'The Climate of Export Credit Agencies'. *Climate Notes*, October. Washington DC: World Resources Institute.

Mielnik, O., and Goldemberg, J. (2001), *Foreign Direct Investment and Decoupling Between Energy and Gross Domestic Product in Developing Countries*. International Energy Initiative/ Instituto de Eletrotécnica e Energia, University of São Paulo.

OECD (1999), *Financial Flows to Developing Countries in 1998. Rise in Aid; Sharp Fall in Private Flows*. News Release. Paris: OECD. 10 June.

Pearce, D. W., Markandya, A. and Barbier, E. B. (1989), *Blueprint for a Green Economy*. London: Earthscan for London Environmental Economics Centre and Department of the Environment.

Siddiqi, T. A. (1990), 'Factors Affecting the Transfer of High Technology to the Developing Countries', in M. Chatterjee, ed., *Technology Transfer to the Developing Countries*. London: Macmillan.

Simmonds, L. and Taylor, P. G. (2000), AEA Technology Environment, *Post Kyoto: Flexibility Mechanisms – A Tool for Implementing the Campaign for Take Off*. Interim report for DGXVII of the European Commission. Didcot: AEAT.

Slesenger, T., with de Bakker, P. and Nikolakaki, S. (2001), *PRESSEA-II*. Final report to the European Commission, DGTREN. DIS-2016–98-GB. Didcot: AEAT.

Smith, A. and Marsh, G. (1997), AEA Technology Environment (ETSU). *Common Actions for Renewable Electricity Generation*. A report for the IEA. Didcot: AEAT.

UN (1993), *Agenda 21: Programme of Action for Sustainable Development*. New York: United Nations.

UN (1999), *The Clean Development Mechanism: Building International Public–Private Partnerships. A Preliminary Examination of Technical, Financial and Institutional Issues*. Unedited draft. October. Based on the deliberations of the ad hoc working group on the CDM, organized jointly by UNCTAD, UNDP, UNEP and UNIDO.

UNCTAD (1996), *World Investment Report 1996: Investment, Trade and International Policy Arrangements*. Geneva: United Nations Conference on Trade and Development.

UNCTAD (1998), *World Investment Report 1997: Transnational Corporations, Market Structure and Competition Policy*. Geneva: United Nations Conference on Trade and Development.

UNFCCC (2000), *Activities Implemented Jointly Under the Pilot Phase. Subsidiary Body for Implementation*. Fourth synthesis report and draft revised uniform reporting format. FCCC/SB/ 2000/6. UN Framework Convention on Climate Change. Available at <www.UNFCCC.org>.